EOS

science strategy for the earth observing system

a r t h

b s e r v i n g

y s t e m

D1540688

Ghassem Asrar
National Aeronautics and Space Administration

Jeff Dozier
University of California—Santa Barbara

Foreword by Charles Kennel
Associate Administrator
Office of Mission to Planet Earth
National Aeronautics and Space Administration

July 1994

Acknowledgments

We extend our thanks to members of the EOS Investigators Working Group and its Science Executive Committee for their contributions in drafting, reviewing, and critiquing earlier versions of this document. Many individuals at Goddard Space Flight Center and NASA Headquarters also deserve our appreciation for providing material and programmatic perspective. Some people warrant mention because of their substantive input and comments:

Mark Abbott, Oregon State University
Wayman Baker, National Weather Service
Bruce Barkstrom, Langley Research Center
Eric Barron, Pennsylvania State University
Peter Brewer, Monterey Bay Aquarium Research Institute
Moustafa Chahine, Jet Propulsion Laboratory
Robert Dickinson, University of Arizona
David Diner, Jet Propulsion Laboratory
James Drummond, University of Toronto
Michael Freilich, Oregon State University
Lee-Lueng Fu, Jet Propulsion Laboratory
John Gille, National Center for Atmospheric Research
James Hansen, Goddard Institute for Space Studies
Dennis Hartmann, University of Washington
Bryan Isacks, Cornell University
Yann Kerr, Centre National de la Recherche Scientifique
Michael King, Goddard Space Flight Center
William Lau, Goddard Space Flight Center
W. Timothy Liu, Jet Propulsion Laboratory
Michael Lizotte, BDM International, Inc.
M. Patrick McCormick, Langley Research Center
Peter Mouginis-Mark, University of Hawaii
Frank Muller-Karger, University of South Florida
Mark Pine, NASA Headquarters
Drew Rothrock, University of Washington
David Schimel, National Center for Atmospheric Research
Mark Schoeberl, Goddard Space Flight Center
Soroosh Sorooshian, University of Arizona
Byron Tapley, University of Texas–Austin
Joe Waters, Jet Propulsion Laboratory
JoBea Way, Jet Propulsion Laboratory
Bruce Wielicki, Langley Research Center
Richard Willson, Jet Propulsion Laboratory
H. Jay Zwally, Goddard Space Flight Center

Special thanks to David Dokken of the Earth Science Support Office for his extensive editorial effort—from initial drafts to final product.

QC
361.75
A85
1994
.1

Foreword

With humanity having grown to envelop the entire globe, and the economic expectations of all nations increasing, comes the realization that we indeed have a responsibility to safeguard the planet—not merely to sustain and enhance our own quality of life, but to ensure a viable world for future generations and the species with whom we share the planet. The early signs of the global influence we exert have already become apparent, as evidenced by the ozone hole, potential global warming, and deforestation, and the constant and deserved attention that these issues receive in the mass media. To preserve our quality of life, we need to counter undesirable global change with responsible adaptation and mitigation strategies. World leaders recognize their obligation, but first the Earth science community must provide the informed guidance needed to formulate environmental policy.

Today, scientists are examining the Earth as a coupled, nonlinear system. Interdisciplinary means are now employed to understand global interactive processes, which in itself introduces uncertainty. Earth system science has subsumed traditional Earth science disciplines and added others, and incomplete understanding in one discipline reverberates throughout in ways peripheral to formerly discipline-specific study. At present, it is difficult (or even impossible) to quantify certain processes at many temporal and spatial scales (e.g., climate-cloud interactions). We need to increase our understanding of all the disparate elements of the Earth system to develop a clearer picture of regional as well as global processes.

Only remote sensing from space can provide the global, repeatable, continuous observations of processes needed to understand the Earth system as a whole. Of course, understanding the Earth as an integrated system involves the full range of Earth science disciplines, satellite and ground-based observations, and sophisticated computer models; however, the synergistic measurements afforded by colocated sensors on satellites cannot be replicated *in situ*. Ultimately, if you are studying global processes, you need global measurements, and that entails remote sensing.

NASA has designed an innovative observing and data management strategy to provide Earth scientists the data needed to conduct intensive process-oriented research. Mission to Planet Earth (MTPE) is the agency's contribution to the U.S. Global Change Research Program (USGCRP), which in turn is embedded within the larger international global change research effort. The Earth Observing System (EOS) is the core of the ambitious MTPE Program. Even

iii

KELLY LIBRARY
EMORY & HENRY COLLEGE
EMORY, VA 24327

though the broader goals of EOS have been scaled back over the years in response to fiscal and other external factors, the program will yield more information on Earth system processes than has ever been achieved previously. The instrument complements have been selected specifically to address key attributes of the Earth system, but enough resiliency has been built into the program to allow researchers and policymakers the leeway to address unexpected manifestations of global change. Instruments on follow-on spacecraft can and will be changed with technology advances and evolving observational requirements.

The 15-year time frame planned for EOS came about after long deliberation. The basic design requirement was to provide a long enough data record (i.e., a complete solar cycle) to discern trends and to help differentiate natural from human-induced perturbations to the Earth system. Even more important is the sustained commitment made by the U.S. Government to participate in and contribute substantially to the international mandate to further understanding of our home planet. Undeniably, the repercussions associated with undetected global change could be immense.

Let me stress here that EOS involves far more than a series of remote-sensing satellites. The EOS Program encompasses a wide range of scientific investigations, observational capabilities, a vast data and information system, and educational activities. Research is ongoing, and EOS-funded investigators have already made substantial contributions to international global change assessments. Though the EOS Program specifically seeks to determine the extent, causes, and regional consequences of global *climate* change, investigators are not constrained to deal solely with the biological and chemical components of the Earth system. Those who designed the program and others who now take part acknowledge that physical evidence can serve as indicators or precursors of socio-economic developments. Shifting patterns in the availability of water, for instance, can have a profound influence on regional agricultural and industrial development.

As we move to a global economy and encounter the environmental impacts of our activities, many difficult decisions will have to be made. To make intelligent choices requires that we understand our Earth system much better than we do at present. NASA's Mission to Planet Earth is committed to timely provision of the scientific knowledge needed by world leaders to formulate sound, equitable environmental policies.

Charles Kennel

Dr. Charles Kennel
Associate Administrator, Office of Mission to Planet Earth
NASA Headquarters

Table of Contents

Section   **Page**

FOREWORD iii

INTRODUCTION 1
   NEED FOR AN EARTH OBSERVING SYSTEM 1
     The Inevitability of Global Change 1
     The U.S. Global Change Research Program .3
   THE EARTH SYSTEM: TWO WATER-COUPLED SUBSYSTEMS 6
     Physical Climate Subsystem 6
     Biogeochemical Cycles 8
     Hydrologic Cycle 9
   CHALLENGE FOR EARTH SYSTEM SCIENCE 10
   NEED FOR INTEGRATED MEASUREMENTS 13
     Synergism and Simultaneity 13
     EOS Science Investigations 17

CLOUDS, RADIATION,
WATER VAPOR, AND PRECIPITATION 21

OCEANS: CIRCULATION,
PRODUCTIVITY, AND AIR-SEA EXCHANGE 31

GREENHOUSE GASES
AND TROPOSPHERIC CHEMISTRY 39

LAND SURFACE: ECOSYSTEMS AND HYDROLOGY 45

ICE SHEETS, POLAR AND
ALPINE GLACIERS, AND SEASONAL SNOW 53

OZONE AND STRATOSPHERIC CHEMISTRY .59

VOLCANOES, DUST STORMS, AND CLIMATE CHANGE 65

**AN OVERVIEW OF INTERNATIONAL EARTH-
OBSERVING CAPABILITIES—PLANNED AND PRESENT** **71**
   SCIENCE PRIORITIES FOR USGCRP 71
   COMPONENTS OF THE EOS PROGRAM 74
      The EOS Satellite Series 75
      EOS Data and Information System 78
      Education 83
   PARTNERS IN THE EOS PROGRAM 84
      Mission to Planet Earth Phase I 84
      The International Earth Observing System 85
   SUMMARY 90

ASSESSING THE IMPACT OF CLIMATE CHANGE **91**
   ACCELERATION OF CLIMATE CHANGE 91
   AWARENESS OF CLIMATE CHANGE 93
   THE GLOBAL PERSPECTIVE AFFORDED BY EOS 94
      Atmospheric Ozone and Ultraviolet Radiation Hazards 96
      Climate Variability and Economic Development 96
      Global Warming and Long-Term Climate Change 98
      Ecosystems and Biodiversity 99
      Social and Economic Implications 102
   ASSESSMENT OF CLIMATE CHANGE 103
   CONCLUSIONS 106

APPENDICES
   A. ACRONYMS 109
   B. BIBLIOGRAPHY 113

Graphics

Page

FIGURES

1. NASA's Mission to Planet Earth Program 4
2. The Earth System—Two Subsystems Connected by the Hydrologic Cycle 7
3. Orbits Required to Establish a Global Perspective 14
4. U.S. Global Change Research Program Science Priorities 73
5. EOS and Earth System Modeling 75
6. EOS Mission Launch Profile 76
7. EOSDIS Architecture 79
8. EOSDIS-Sponsored Data Centers 81
9. The Scientific Contribution to Policy Formulation 92
10. EOS—A User-Driven Program 107

TABLES

1. The EOS Satellite Series 12
2. EOS Instruments 16
3. EOS Interdisciplinary Science Investigations 18
4. Japanese Instruments Planned as Part of IEOS 87
5. European Instruments Planned as Part of IEOS 89
6. Atmospheric Ozone and Ultraviolet Radiation Hazards 97
7. Climate Variability 98
8. Greenhouse Warming and Long-Term Climate Change 100
9. Ecosystems and Biodiversity 101
10. Social and Economic Implications of Global Climate Change 103

CHAPTER 1

Introduction

From the beginning of scientific inquiry, humankind has sought a greater knowledge of the Earth. Yet it is only near the close of the 20th century that we have begun to understand the interconnected unity of our planet and have become able to probe its many intricate processes on a global scale.

Earth System Science: A Closer View, 1988

NEED FOR AN EARTH OBSERVING SYSTEM

Since space exploration began in the late 1950s, space technology has enabled investigations of the other planets of the solar system. We have exploited the opportunity that satellites provide to learn about worlds far distant from our own. Now, that same technology provides the global perspective needed for an integrated, long-term, scientific investigation of our home planet.

The Inevitability of Global Change

We know from the geologic record that the Earth has undergone significant alterations in its physical configuration since the planet's formation some 4.5 billion years ago. Indeed, the entire history of the Earth is a dynamic continuum, with changes on all time scales. The evolving character of the Earth makes its scientific study both interesting and difficult. What is disturbing, however, is evidence that the rates of change for several key components of the system are increasing dramatically, and that human activity is responsible. Within the span of a single human lifetime, we can detect significant differences in atmospheric chemistry, terrestrial and oceanic ecosystems, and water quality. Whether human society affects the Earth system is no longer a question. Instead, the questions are, "How severe is that effect? What can be done about it?"

1

Almost 6 billion humans will populate the Earth by the year 2000. Can we anticipate the impact that our burgeoning population will have on the Earth's physical makeup? For example, we know that combustion of fossil fuels pumps more than 5 billion metric tons of carbon into the atmosphere each year—as carbon dioxide, a key radiatively active gas. Deforestation accounts for an additional 1.5 to 2 billion tons, so humanity's contribution is about 7 billion tons above the natural carbon flux. Our measurements of atmospheric carbon dioxide show an annual increase of 3.5 billion tons of carbon, and the ocean takes up perhaps an extra 2 billion tons above the natural flux. Hence we cannot account for more than 1 billion tons of carbon! Where does it go? How will the carbon budget change as atmospheric concentrations of carbon dioxide increase?

We know that at least 11.3 million hectares of mature forest are destroyed annually. What are the repercussions? The human species' voracious appetite for an enhanced quality of life mandates that we mine and mill the Earth's resources. Scientists need to determine how such consumption affects the climate system. Human activities will induce climate change that could cause substantial physical and financial upheaval on a global scale.

Continued change in the Earth system is inevitable. Without scientific evidence documenting the consequences of human action and the variability of natural systems, the effect of increasing radiatively active gas concentrations remains the subject of debate. Decisionmakers need scientific guidance to develop sound policy, particularly for situations where policy choices may involve altering human behavior to protect the environment. Without a firm understanding of how we affect the world, we cannot discriminate between naiveté and overzealous regulation:

- The abundance of fossil fuels and the strong association between standard of living and energy consumption suggest that carbon dioxide emissions into the atmosphere will increase for at least a century. Even if the countries of the world agreed today to freeze emissions of carbon at current rates, the atmospheric concentration would continue to increase. Stabilizing atmospheric carbon dioxide at the current level, which is about 30 percent above the pre-industrial concentration, would require a drastic 70 to 80 percent reduction in emissions.
- World population increases have caused changes in land use and agricultural production in ways that contribute to global change. For example, rice paddies and cattle have become significant sources of methane and other important greenhouse gases in the atmosphere.
- Chlorofluorocarbons (CFCs) manufactured for refrigeration, air conditioning, and industrial applications break down when they reach the stratosphere to produce the chlorine radical that destroys ozone. Because

CFCs have a 100-year residence time in the troposphere, even policies to completely phase out CFC production by the end of this decade will still allow stratospheric chlorine to increase until about 2015. Hence stratospheric ozone will continue to decrease for at least 20 years, and probably will not rise to pre-1970 values until near the end of the 21st century.

The societal implications of human-induced global change could be enormous. Bubbles trapped in ice cores from Antarctica and Greenland show that pre-industrial atmospheric carbon dioxide concentrations were stable for at least 160,000 years. Global climate models predict that a doubling of carbon dioxide concentrations, compared to these pre-industrial values, will result in a globally averaged warming of 2 to 4°C in the troposphere. The last global temperature change that large—a decrease instead of an increase—occurred during the Ice Age advance, and brought continental glaciers inside the boundaries of what is now the coterminous United States.

Global warming is not the only consequence of changing atmospheric composition. The altered circulation associated with shifts in atmospheric temperature will also affect the distribution of rain and snow. Some climate simulations show an increased frequency of drought in mid-latitude continental regions. Such droughts would reduce water availability and quality for industry, agriculture, and human consumption. Warming may cause the sea level to rise, threatening coastal cities and environments. In regions that depend on winter snowfall for their water supply, rain may fall in winter instead of snow. Because it would run off in the winter, the water would be unavailable for agriculture in the spring and summer.

Unfortunately, future global changes probably will not be limited to the causal factors scientists have already identified. Subtle, yet significant, changes that are now undetected are likely to occur. The inevitability of global change and the prospect of unanticipated modifications of the global environment add urgency to the need to understand the operation of the Earth system.

The U.S. Global Change Research Program

Worldwide recognition of global change as an issue affecting all the Earth's inhabitants has led to a major international research collaboration. The United States participates through the U.S. Global Change Research Program (USGCRP), which coordinates the efforts of numerous Federal agencies under the guidance of the Committee on the Environment and Natural Resources (CENR). The National Aeronautics and Space Administration (NASA) contributes via its Mission to Planet Earth (MTPE) Program (see Figure 1). The

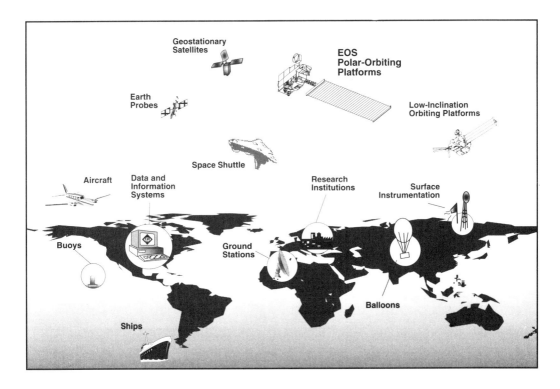

FIGURE 1. NASA's Mission to Planet Earth Program

Earth Observing System (EOS) serves as the foundation of MTPE by providing instrument and platform hardware, a community of funded scientists, and the infrastructure to consolidate data and information from surface campaigns and remote-sensing satellites.

The MTPE Program employs a space- and surface-based measurement strategy to provide the scientific basis to address global change. Interagency and international collaborations (i.e., MTPE Phase I) precede the EOS flight segment, establishing and continuing the baseline data sets needed to document trends. In addition to the EOS and MTPE Phase I satellites sponsored by NASA, the polar-orbiting and mid-inclination platforms from Europe, Japan, and the U.S. National Oceanic and Atmospheric Administration (NOAA) round out an International Earth Observing System (IEOS).* These international satellite contributions employ proven technologies, with all instruments sporting a heritage of successful spaceflight. Early remote-sensing experiments in space settled for low resolution and accuracy to break new ground. Because the scientific requirements have since become more refined, better instrumentation and modeling are now needed.

---

*See Chapter 9 for a brief overview of the missions that comprise IEOS.

Why is EOS so important? Today, we are almost completely ignorant of how the Earth "works" as a system. The EOS Program is the first integrative, large-scale initiative to explore the functioning of the planet as a whole. In addition to the broader goal of learning about the Earth as part of the continuing scientific enterprise, the need to understand human-induced changes drives this mission. Many types of measurements must be collected over an extended time period to differentiate natural variability from human perturbations. Developing a comprehensive data record on myriad levels requires a sustained commitment. Snapshots do not provide the perspective needed to determine trends or to quantify magnitudes of change. Certainly, short-term missions and regional or local campaigns that use aircraft remote sensing identify and elucidate phenomena that indicate global change. However, the spatial, temporal, and spectral coverage offered by modern space-based instrumentation best addresses the requirements of researchers for long time series of data. Consistent, well-calibrated observations from space enhance the valuable work in the field and the laboratory. The EOS satellite series will operate in a variety of orbits for at least 15 years to help distinguish between natural variability and human-induced changes.

With the revelation that the Earth is a vast and complex system not easily subdivided into specific components, interdisciplinary studies have moved to the fore. Interdisciplinary research weaves elements from such formerly disparate sciences as terrestrial ecology, oceanography, and climatology. Process studies blur traditional boundaries between the sciences and introduce even more challenges for satellite and instrument developers. Parameters formerly of interest to just one discipline now have applications in multiple others. Earth science from space goes beyond extracting threads of information from a single sensor; rather, processed data from a bevy of instruments are fused input to or compared with models. Data systems thus need to make the information more readily available and usable. The Earth is a complicated realm, and to delve into its mystery requires enlightened scientific expertise and dedication in all facets of the global change research effort—observational capabilities, data and information systems, scientific investigations, and education.

This document focuses on the science sought during the EOS observational period and beyond. If researchers can improve predictions of the plausible consequences of future climate change, society will have the chance to adapt to and mitigate negative effects. Only through scientific study can we hope to better understand Earth as a system. Space-based monitoring alone is not sufficient. The scientists who develop the instruments and algorithms and who build and refine Earth system models are the ones who explore, anticipate, and ultimately predict global climate change. EOS and related programs merely provide the resources to accomplish these tasks.

THE EARTH SYSTEM:
TWO WATER-COUPLED SUBSYSTEMS

For our presentation of the EOS observational strategy, we must first describe the essential features of the Earth system. The Earth system is complex, and important manifestations of global change occur in all its components over a range of spatial and temporal scales.

Earth is unique among the planets in its abundance of water in all three phases—gas, liquid, and solid—a consequence of the Earth's radiative balance. In turn, this balance is strongly affected by global water cycles and biogeochemistry. Another unique feature of the Earth is that oxygen and anti-oxidant gases such as methane coexist in the atmosphere; the consequent disequilibrium for long periods means that atmospheric composition and climate are intimately tied to biological processes like photosynthesis and decomposition. Therefore, understanding even the simplest aspects of the Earth system requires knowledge of geophysics, geochemistry, and biology.

We regard the Earth system as two subsystems—physical climate and biogeochemical cycles—linked by the global hydrologic cycle (see Figure 2). Examination of these subsystems and their linkages defines the critical questions that EOS addresses.

Physical Climate Subsystem

The physical climate subsystem is sensitive to fluctuations in the radiation balance of the Earth. These modifications result from increasing atmospheric concentrations of radiation-absorbing gases and from changes in planetary albedo. Moreover, it now is apparent that the perturbations to the planet's radiative heating mechanism caused by human activities rival or exceed natural change.

According to the Intergovernmental Panel on Climate Change (IPCC), increases in radiatively active greenhouse gases (primarily carbon dioxide, methane, and CFCs) between 1765 and 1990 have caused a radiative forcing of 2.5 Wm^{-2} (Houghton et al., 1992). This increased trapping of infrared radiation by the Earth's atmosphere represents about 1 percent of the globally averaged emission of infrared radiation by the Earth system. Although this would appear to be a small amount, if there is no policy response to this trend, emissions of greenhouse gases will cause a rate of increase of global mean temperature during the next century of about 0.2 to 0.5°C per decade. This range in the estimate of global warming rates

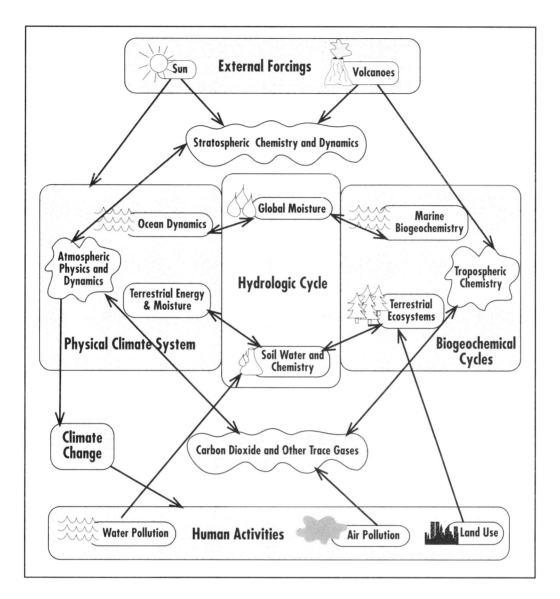

FIGURE 2. The Earth System—Two Subsystems Connected by the Hydrologic Cycle

reveals the uncertainty surrounding global climate models, caused primarily by knowledge gaps in quantifying cloud generation and radiative processes.

Not all expected warming will occur immediately, because of the enormous heat-absorbing capacity of the oceans. Mesoscale and small-scale processes play a critical role in the transfer of heat from the atmosphere into the oceans. Present models do not yet resolve these processes, but we know that there is a strong link between mesoscale ocean eddies and vertical mixing. Turbulent mixing of the added heat within the top 50 to 200 m of the ocean and vertical mixing to the deeper layers may delay warming of the atmosphere.

The need for accurate observation, modeling, and prediction of the effects of anthropogenic changes to the Earth's climate leads to the following specific questions that need to be answered if we are to prepare for future conditions:

- How might the roles of clouds, water vapor, and aerosols in the Earth's radiation and heat budgets change with increased atmospheric greenhouse gas concentrations? What other radiative feedback mechanisms lead or respond to changes in climate?
- How do the oceans interact with the atmosphere in the transport and uptake of heat, and what role do they play in global and regional climatic variability?
- How do atmospheric circulation and oceanic circulation affect the mass balance of ice sheets, glaciers, and ice caps, and what is the effect on sea level?
- How do land-surface properties (e.g., snow cover, evapotranspiration, land use, vegetation) influence circulation?
- How do changes in the energy output of the Sun influence climate?

Answering these questions alone is a challenge, but they are only the beginning of the list, which lengthens with a cursory examination of the other major subsystem and interactions through the water cycle.

Biogeochemical Cycles

In the past 100 years, human actions have altered fundamental elements of the biogeochemical cycles of the Earth. Since 1850, atmospheric carbon dioxide has increased by about 30 percent, atmospheric methane by more than 100 percent. It also appears that the capacity of the troposphere to oxidize trace constituents has decreased during the same period, which could lead to more rapid accumulation of certain gases.

Other changes in atmospheric chemistry have occurred. Ozone concentrations in the stratosphere have decreased, most notably above Antarctica and the Southern Ocean, and in the mid-latitudes of the northern hemisphere. Increased levels of ultraviolet radiation reach the Earth's surface as stratospheric ozone concentrations decrease, and this ultraviolet radiation can harm living organisms. On the other hand, both rural and urban areas worldwide have experienced increases in tropospheric ozone—a pollutant that also poses serious health hazards.

Changes in land use and management, water pollution, and the changing atmosphere have affected cycles of essential elements within ecosystems. These human impacts have influenced interactions between terrestrial and marine ecosystems, the atmosphere, and the hydrosphere. Precipitation has become acidic over widespread areas, agriculture and forestry practices have increased rates of soil erosion, and land use and industrial development continue to alter the chemical composition of surface water, groundwater, and coastal seawater.

Global change researchers must address the following basic questions to shed light on the Earth's changing biogeochemistry:

- What roles do the oceanic and terrestrial components of the biosphere play in the changing global carbon budget? How are these roles shifting as atmospheric carbon dioxide increases?
- How do changes in the physical/chemical climate system link to changes in photosynthesis and evaporation? What are the principal natural and anthropogenic sources of carbon dioxide and methane, and what are the contributions from fossil fuels, biomass burning, and changes in land use versus natural sources?
- Is the changing chemical composition of the troposphere altering the atmospheric lifetimes of key radiatively and chemically active gases? What determines the oxidizing capacity of the troposphere, and how do biological and industrial processes affect it?
- What are the likely effects on natural and managed ecosystems of increased carbon dioxide, acid deposition, shifting patterns of precipitation, and changes in soil erosion, river chemistry, and atmospheric ozone concentrations?
- What are the feedback mechanisms and linkages between climate, atmospheric and river-borne geochemical loading to the ocean, and ecosystems?
- What are the mechanisms that determine magnitudes, locations, and frequencies of volcanic eruptions, and what are their effects on climate? How do changing patterns of land use affect the flux of aerosols and wind-borne dust into the atmosphere, and how might these processes affect the climate system?

Hydrologic Cycle

The hydrologic cycle links the physical climate and biogeochemical cycles. The phase change of water between its gaseous, liquid, and solid states involves storage and release of latent heat, so it influences atmospheric circulation and globally redistributes both water and heat. Water's efficiency as a solvent also proves critical to biogeochemical processes. Thus, the hydrologic cycle is the integrating process for the fluxes of water, energy, and chemical elements among components of the Earth system (National Research Council, 1991).

To understand how global change may influence global water balances, we must know more about spatial and temporal variations in the storage of water in its various reservoirs and the magnitude of transfers between these reservoirs. In addition, the regulation of these water transfers by physical and biological processes and their influence on climate need more detailed investigation. Specific questions about the hydrologic cycle for the Earth science community include the following:

- How will atmospheric variability, human activities, and climate change affect patterns of humidity, precipitation, evapotranspiration, and soil moisture—as well as distributions of liquid water, snow, and ice—on the Earth's surface? How would any redistribution influence global sea level?
- What controls the mechanisms for the transfer of water among the hydrologic reservoirs, and what are the associated directions and magnitudes of transfer?
- How does soil moisture vary in time and space, and what are the mechanisms controlling rainfall in arid and semi-arid regions? How does this variability affect the hydrology, geomorphology, and ecology of such regions, and how do these variations translate into regional fields of carbon exchange?
- What is the role of climatic change in the cryosphere? The cryosphere has an internal system and a margin in which strong interactions occur with the physical and biogeochemical subsystems. How will the cryospheric system change internally, and how will these changes ripple through the global climate system?
- How well can we predict changes in the global hydrologic cycle using present and future observation systems and models? How can we best determine the interannual variability of global hydrologic processes, from natural variability and the seasonal cycle, to infer the mechanisms and magnitudes of climate change?

These questions, and many more, require answers before we can hope to predict and adjust to future global changes in the Earth system. EOS is designed specifically to provide the opportunity for scientists and policymakers to address these questions. Chapters 2 through 8 focus on the specific answers to be sought, via space-based observations and scientific investigations currently underway and continuing throughout the lifetime of the EOS Program.

CHALLENGE FOR EARTH SYSTEM SCIENCE

In addition to integrated measurements of Earth system processes, researchers must also adopt an interdisciplinary approach in analyzing the collected data and disseminating the resulting information. In the past, the diverse disciplines that comprise the Earth sciences developed independently, and scientists and engineers tended to pursue discrete research objectives and strategies. Advances in observational methods, theories, and models in the fields of meteorology, oceanography, climatology, ecology, geochemistry, geomorphology, and hydrology remained unique. Now, however, three forces have combined to alter the modes and focus of research:

1) Global change research has reached the point where progress requires an integration of traditional disciplines, and many disciplines have reached the state of maturity where they are ready for that integration.
2) The view of Earth from space has underscored the fact that the planet is a single, complex, integrated system.
3) The growing awareness and apprehension about the effects of human-induced global change make a concerted scientific effort essential, and they increase the importance of reanalyzing remote-sensing data acquired over the past 2 decades to understand and predict the evolution of the Earth system and our role in it.

These three forces have led to the definition of several central problems that require the unified perspective now known as "Earth System Science" (Earth System Sciences Committee, 1988):

- The greenhouse effect associated with increasing concentrations of carbon dioxide, methane, CFCs, and other radiatively active gases
- Ozone depletion in the stratosphere, resulting in a significant increase in ultraviolet radiation reaching the Earth's surface
- A diminishing supply of water suitable for human uses
- Deforestation and other anthropogenic changes to the Earth's surface, potentially affecting the carbon budget, patterns of evaporation, precipitation, soil erosion, and other components of the system
- Changes in photosynthesis, respiration, transpiration, and trace gas exchange both on the land and in the ocean
- Decline in the health of vegetation caused by long-term changes in the chemistry of the atmosphere, precipitation, runoff, and groundwater.

For those who make observations of the Earth system and develop models of its operation, Earth system science means the creation of interdisciplinary models that couple elements from formerly disparate sciences. This approach mandates a broad remote-sensing strategy to ensure collection of the data required by the entire scientific community. EOS satisfies the dual requirements of an integrated approach to Earth observation and the development of meaningful geophysical data products to aid global change investigations.

The EOS Program, in concert with its international counterparts, addresses high-priority science and environmental policy issues in Earth system science by initiating and funding interdisciplinary teams of scientists, by developing the EOS Data and Information System (EOSDIS), and by preparing for present and future remote-sensing missions (see Table 1). The data and information system provides access to vast stores of remote-sensing data, and allows Earth scientists to study processes and to develop better models to assess the impacts of global climate change. Experience has demonstrated the need for a functional data system to acquire, process, manage, and distribute Earth science data, particularly for satellite remote-sensing data. Having a comprehensive data and information system is critical to mission success, considering the wealth and diversity of data and information from the EOS satellite series, international operational missions, the research and development missions of our international and interagency partners, and surface campaigns. All these data need to be processed, archived, and made readily available.

TABLE 1. The EOS Satellite Series

Satellites (First Launch)	Mission Objectives
EOS-AM Series (1998) Earth Observing System Morning Crossing (Descending)	Clouds, aerosols and radiation balance, characterization of the terrestrial ecosystem; land use, soils, terrestrial energy/moisture, tropospheric chemical composition; contribution of volcanoes to climate; and ocean primary productivity (includes Canadian and Japanese instruments)
EOS-COLOR (1998) EOS Ocean Color Satellite	Ocean primary productivity
EOS-AERO Series (2000) EOS Aerosol Mission	Distribution of aerosols and greenhouse gases in the stratosphere and upper troposphere (spacecraft to be provided through international cooperation)
EOS-PM Series (2001) Earth Observing System Afternoon Crossing (Ascending)	Cloud formation, precipitation, and radiative properties; atmospheric temperature and moisture profiles; air-sea fluxes of energy and moisture; sea-ice extent; and soil moisture and snow over land (includes European instruments)
EOS-ALT Series (2002) EOS Altimetry Mission	Ocean circulation and ice-sheet mass balance (includes French instruments)
EOS-CHEM Series (2003) EOS Chemistry Mission	Atmospheric chemical composition and dynamics; chemistry-climate interactions; air-sea exchange of chemicals and energy (to include an as-yet-to-be-determined Japanese instrument)

Chapter 9 offers concise coverage of the EOS Program—both the implementation strategy and its various components. The *EOS Reference Handbook* (Asrar and Dokken, 1993) gives additional details.

NEED FOR INTEGRATED MEASUREMENTS

Synergism and Simultaneity

Predictions of global change show enormous potential for human activity to have climatic, biological, and hydrological consequences. Unfortunately, given the critical observational limitations researchers now face, even the most comprehensive models must hedge answers with large uncertainties. EOS will monitor environmental changes to advance understanding of the entire Earth system—developing a deeper comprehension of the components of that system and the interactions among them. To quantify changes in the Earth system, EOS will provide systematic, continuous observations from low Earth orbit for a minimum of 15 years. By enhancing understanding of the processes involved, EOS will help discriminate between anthropogenic and natural changes.

Scientists need long-term, consistent measurements of the key physical variables that define the shifts in state and variability of Earth system components—the atmosphere, hydrosphere, cryosphere, oceans, and land surface. Lacking these measurements, predictions of the complex responses of the Earth system to human activities and natural variations lack an adequate baseline to determine trends. Surface and airborne measurement campaigns are intrinsic to the observational strategy. By design, field experiments and airborne remote-sensing campaigns observe the smaller spatial and temporal scales necessary to validate the accuracy of the geophysical and biological parameters remotely sensed from space. Neither *in situ* nor space-based instruments can measure all physical processes active in the Earth system; however, space-based observations hold the key, because satellites best capture a consistent, global perspective.

To date, remote sensing of the Earth system has consisted largely of discipline-oriented missions focused on a narrow range of physical phenomena and problems. While the current approach has led to significant gains in understanding the atmosphere, hydrosphere, and biosphere, it does not provide enough information about the coupling of these systems. Data from this discipline-specific approach cannot quantify the fluxes of mass (i.e., water, carbon dioxide, and other trace constituents), heat, and momentum between the land, atmosphere, and oceans. Current computer simulations include only primitive models of these interactions, which drive important changes in the Earth system.

Determination of the magnitudes and spatial variations of global changes requires consistent global measurements, over a long enough period to separate long-term trends from natural variability associated with seasons and other cyclical or periodic events. The duration of the observations must be at least 15 years, to span a solar sunspot cycle and several El Niño events. These observations must characterize the whole planet and its regional variations and enable quantification of the processes that govern the Earth system.

The full set of observations requires different instruments on satellites in different orbits (see Figure 3). No single orbit permits the gathering of all the needed information on Earth processes. To obtain global coverage, satellites must fly in both polar and inclined, equatorial or near-equatorial, orbits—polar orbits to view the entire planet, and inclined orbits to sample the diurnal cycle at lower latitudes. In the next century, NASA plans a constellation of geostationary satellites, which will resolve dynamic processes that operate on the scale of minutes to hours. Geostationary platforms can observe the full Earth disk, thereby revealing unpredictable short-term events and obtaining weak signals that only instruments capable of "staring" for relatively long periods of time can detect.

Global change researchers meld data from multiple instruments and disparate disciplines. The EOS science strategy considers two types of synergism—instruments and science. Instrument synergism implies that instruments must be grouped

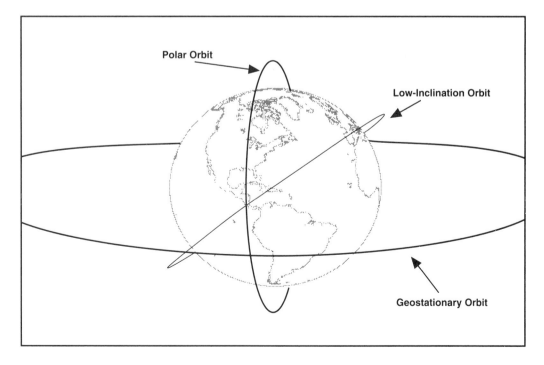

FIGURE 3. Orbits Required to Establish a Global Perspective

together. Measurements by one instrument are required to interpret those of another, so the measurements cannot be separated too widely without compromising the utility of some of the observations. Science synergism means that groups of instruments measure 1) different components of a single process; 2) similar components of different, but correlated, processes; and/or 3) coupled processes that control the functioning of the Earth system. Certain instruments must acquire near-simultaneous data to satisfy scientific objectives. Examples include the study of clouds and radiation forcing and feedback, measurements of the sources and sinks of greenhouse gases, and measurements of atmospheric, oceanic, and terrestrial components of the hydrological and biogeochemical cycles.

Simultaneity plays a key role in the EOS observational strategy. Understanding the Earth as a system requires measurements of many variables over the whole Earth at appropriate spatial and temporal scales. Many of these measurements and interpretations require simultaneous observations of coupled phenomena. For example, estimating photosynthesis rates requires measurements of plant biomass, incident sunlight, soil moisture, and air temperature. If researchers hope to capture this process in a model, an instrument complement that can measure these different parameters would have to acquire near-simultaneous data to yield a complete, optimized picture.

Within the full EOS instrument suite (see Table 2), sets of instruments complement one another to produce desired observations. In some instances, certain instrument suites must generate correlative data; therefore, they must make their measurements simultaneously. For example, measurement of ocean circulation requires global altimeter data, precise orbit determination, and water vapor amounts. These observations also contain signals resulting from ocean tide deformation, atmospheric pressure and ocean surface interactions, and sea-level change, including long-term average sea level as well as interannual hemispherical variations. Such sets of observations dictate some minimum payload groupings. Chapters 2 through 8 provide more detail on these requirements.

The time scales for significant variations in the phenomena being observed define the meaning of "simultaneous." Three principal time scales prove important to this aspect of the observing strategy:

- **Time Scales of Years**—To define the subtle, but critical, interactions between processes that modulate and regulate climate, researchers need long time series of data to differentiate secular trends from natural variability. For this reason, the EOS framework emphasizes continuous, long-term measurements.

TABLE 2. EOS Instruments

Instruments in the Early EOS Period (1997-2001)

❏ **Atmospheric Infrared Sounder/Advanced Microwave Sounding Unit/Microwave Humidity Sounder (AIRS/AMSU/MHS)**—Synergistic package that provides temperature and humidity sounding with much better accuracy than current sensors, with temperatures measured to 1K accuracy at 1-km vertical resolution
Team Leader—Moustafa Chahine, Jet Propulsion Laboratory

❏ **Advanced Spaceborne Thermal Emission and Reflection Radiometer (ASTER)**—Yields high-resolution images of land surface, water, and clouds from visible through thermal infrared wavelengths; one stereophotogrammetric band to enable production of digital elevation models
Team Leaders—Hiroji Tsu, Geological Survey (Japan) and Anne Kahle, Jet Propulsion Laboratory

❏ **Clouds and Earth's Radiant Energy System (CERES)**—Flies on multiple satellites in morning, afternoon, and inclined orbits to measure Earth's radiation balance
Principal Investigator—Bruce Barkstrom, NASA Langley Research Center

❏ **EOS Ocean Color Satellite (EOS-COLOR)**—Observes afternoon ocean color to provide continuity of SeaWiFS measurements of biological production; data purchase mission involving a single launch
Principal Investigator—To be chosen by NASA

❏ **Lightning Imaging Sensor (LIS)**—Flies in an inclined orbit on TRMM in 1997, providing global observations of lightning distribution and its variability
Principal Investigator—Hugh Christian, NASA Marshall Space Flight Center

❏ **Multifrequency Imaging Microwave Radiometer (MIMR)**—Measures precipitation, cloud water, sea-surface temperature and winds, snow and ice extent, snow water equivalence, and soil moisture
Team Leaders—Roy Spencer, NASA Marshall Space Flight Center and an international counterpart to be chosen by ESA

❏ **Multi-Angle Imaging SpectroRadiometer (MISR)**—Globally measures aerosol characteristics, bidirectional reflectances, and albedos at the top of the atmosphere and surface, cloud-top elevation and distribution, and vegetation properties; provides moderate-resolution topographic data through stereophotogrammetry
Principal Investigator—David Diner, Jet Propulsion Laboratory

❏ **Moderate-Resolution Imaging Spectroradiometer (MODIS)**—Flies in both morning and afternoon orbits to discern cloud cover, vegetation, ocean color, surface temperatures, aerosols, and other global geophysical and biological processes for the land, ocean, and atmosphere
Team Leader—Vincent Salomonson, NASA Goddard Space Flight Center

❏ **Measurements of Pollution in the Troposphere (MOPITT)**—Measures carbon monoxide and methane
Principal Investigator—James Drummond, University of Toronto (Canada)

❏ **SeaWinds (NSCAT II)**—Slated to fly on ADEOS II in 1999, provides all-weather measurements of near-surface wind velocity and horizontal stress over the oceans
Principal Investigator—Michael Freilich, Oregon State University

❏ **Stratospheric Aerosol and Gas Experiment III (SAGE III)**—Generates global profiles of aerosols, clouds, ozone and related species, temperature, and pressure in the stratosphere
Principal Investigator—M. Patrick McCormick, NASA Langley Research Center

- **Time Scales of a Few Days**—Some EOS instruments need to be in orbit at the same time, to enable many of the variable components of the Earth system to be characterized globally every 1 to 3 days. While many aspects of the Earth do not change on the time scale of days, all EOS instruments observe dynamic phenomena that do change rapidly.

TABLE 2. EOS Instruments (Continued)

Additional EOS Instruments (Beyond 2001)

- **Active Cavity Radiometer Irradiance Monitor (ACRIM)**—Measures solar irradiance reaching Earth
  Principal Investigator—Richard Willson, Jet Propulsion Laboratory
- **Doppler Orbitography and Radiopositioning Integrated by Satellite/Solid-State Altimeter/TOPEX Microwave Radiometer (DORIS/SSALT/TMR)**—Synergistic package that provides ocean wave height and surface current velocity, sea-surface topography, wind speed, and atmospheric water vapor profiles
  Team Leaders—To be determined by NASA and CNES
- **Earth Observing Scanning Polarimeter (EOSP)**—Globally maps radiance and linear polarization of reflected sunlight to infer aerosol characteristics
  Principal Investigator—Larry Travis, NASA Goddard Institute for Space Studies
- **Geoscience Laser Altimeter System (GLAS)**—Measures the topography of land, glaciers, and ice sheets, and cloud and aerosol layer heights and thickness
  Team Leader—Bob Schutz, University of Texas–Austin
- **High-Resolution Dynamics Limb Sounder (HIRDLS)**—Globally measures temperatures, water vapor, and chemical species in the upper troposphere and stratosphere
  Co-Principal Investigators—John Barnett, Oxford University and John Gille, National Center for Atmospheric Research
- **Microwave Limb Sounder (MLS)**—Globally measures parameters essential for assessing ozone depletion (radicals, reservoirs, source gases) and climate change (upper tropospheric water vapor and other greenhouse gases), with quality not degraded by aerosols or clouds
  Principal Investigator—Joe Waters, Jet Propulsion Laboratory
- **Solar Stellar Irradiance Comparison Experiment II (SOLSTICE II)**—Measures ultraviolet solar irradiance
  Principal Investigator—Gary Rottman, National Center for Atmospheric Research
- **Tropospheric Emission Spectrometer (TES)**—Provides global, three-dimensional profiles of virtually all infrared-active gases from the Earth's surface to the lower stratosphere
  Principal Investigator—Reinhard Beer, Jet Propulsion Laboratory

- **Time Scales of a Few Minutes**—Changes in temperature, aerosol concentrations, water vapor distributions, and clouds can significantly affect the atmospheric contribution to signals received by optical instruments. This imposes simultaneity requirements on separate sets of instruments to make their observations within a few minutes of one another and from the same perspective.

EOS Science Investigations

Planning for the EOS mission began in the early 1980s. In 1988, NASA issued an Announcement of Opportunity (AO) for the selection of instruments and science teams; 458 proposals were received in response to the AO. Early in 1990, NASA announced the selection of 30 instruments, along with their science teams, and 29 Interdisciplinary Science (IDS) investigations. As the program has evolved and as the budget has declined, the number of instruments has been reduced from 30 to 19. The instrument and IDS teams were selected to conduct basic research, to develop methods and models for analysis of EOS observations, to develop and refine models of Earth system processes, and to forge new alliances among scientific disciplines, fostering a unique perspective into how the Earth functions as an integrated system. Table 2 identifies the instruments and their team leaders; Table 3 identifies the IDS teams and their principal investigators. The teams for the six internationally contributed EOS instruments are co-sponsored by NASA and the home institutions of the non-U.S. team members. Some IDS teams have as few as five co-investigators while others have more than 20.

Many of the important scientific questions to be studied by EOS require analyses by interdisciplinary teams using data and information from multiple EOS and IEOS instruments and *in situ* observations. The IDS and instrument teams bring together highly talented experts from diverse fields to tackle specific areas of uncertainty regarding the functioning of the Earth as a coupled system. All investigations will exploit the newly

TABLE 3. EOS IDS Investigations

Coupled Atmosphere-Ocean Processes and Primary Production in the Southern Ocean
Mark Abbott, Oregon State University

Global Water Cycle: Extension Across the Earth Sciences
Eric Barron, Pennsylvania State University

Long-Term Monitoring of the Amazon Ecosystems through EOS: From Patterns to Processes
Getulio Batista, Instituto Nacional de Pesquisas Espaciais (Brazil)
Jeffrey Richey, University of Washington

Biogeochemical Fluxes at the Ocean/Atmosphere Interface
Peter Brewer, Monterey Bay Aquarium Research Institute

Northern Biosphere Observation and Modeling Experiment
Josef Cihlar, Canada Centre for Remote Sensing (Canada)

NCAR Project to Interface Modeling on Global and Regional Scales with EOS Observations
Robert Dickinson, University of Arizona

Hydrology, Hydrochemical Modeling, and Remote Sensing in Seasonally Snow-Covered Alpine Drainage Basins
Jeff Dozier, University of California–Santa Barbara

Interdisciplinary Studies of the Relationships between Climate, Ocean Circulation, Biological Processes, and Renewable Marine Resources
J. Stuart Godfrey, CSIRO (Australia)

Use of a Cryospheric System (CRYSYS) to Monitor Global Change in Canada
Barry Goodison, Canada Centre for Remote Sensing (Canada)

Observational and Modeling Studies of Radiative, Chemical, and Dynamical Interactions in the Earth's Atmosphere
William Grose, NASA Langley Research Center

Interannual Variability of the Global Carbon, Energy, and Hydrologic Cycles
James Hansen, NASA Goddard Institute for Space Studies

Climate Processes Over the Oceans
Dennis Hartmann, University of Washington

Climate, Erosion, and Tectonics in the Andes and Other Mountain Systems
Bryan Isacks, Cornell University

The Hydrologic Cycle and Climatic Processes in Arid and Semi-Arid Lands
Yann Kerr, LERTS (France)
Soroosh Sorooshian, University of Arizona

TABLE 3. EOS IDS Investigations (Continued)

Global Hydrologic Processes and Climate
William Lau, NASA Goddard Space Flight Center

**The Processing, Evaluation, and Impact
on Numerical Weather Prediction of AIRS, AMSU, and
MODIS Data in the Tropics and Southern Hemisphere**
John LeMarshall, Bureau of Meteorology Research Centre (Australia)

**The Role of Air-Sea Exchanges
and Ocean Circulation in Climate Variability**
W. Timothy Liu, Jet Propulsion Laboratory

Changes in Biogeochemical Cycles
Berrien Moore III, University of New Hampshire

**A Global Assessment of Active
Volcanism, Volcanic Hazards, and Volcanic
Inputs to the Atmosphere from EOS**
Peter Mouginis-Mark, University of Hawaii

**Investigation of the Atmosphere-Ocean-Land
System Related to Climatic Processes**
Masato Murakami, Japan Meteorological Agency (Japan)

**Chemical, Dynamical, and Radiative Interactions
through the Middle Atmosphere and Thermosphere**
John Pyle, Cambridge University (United Kingdom)

**The Development and Use of a
Four-Dimensional Atmospheric-Ocean-Land
Data Assimilation System for EOS**
Richard Rood, NASA Goddard Space Flight Center

**Polar Exchange at the Sea Surface (POLES):
The Interaction of Oceans, Ice, and Atmosphere**
Drew Rothrock, University of Washington

**Using Multi-Sensor Data to Model Factors
Limiting Carbon Balance in Global Grasslands**
David Schimel, NCAR and Colorado State University

**Investigation of the Chemical and
Dynamical Changes in the Stratosphere
Up to and During the EOS Observing Period**
Mark Schoeberl, NASA Goddard Space Flight Center

Biosphere-Atmosphere Interactions
Piers Sellers, NASA Goddard Space Flight Center

Middle and High Latitudes Oceanic Variability Study
Meric Srokosz, British National Space Centre (United Kingdom)

**Earth System Dynamics:
The Determination and Interpretation of
the Global Angular Momentum Budget Using EOS**
Byron Tapley, University of Texas–Austin

**An Interdisciplinary Investigation of Clouds
and Earth's Radiant Energy System: Analysis**
Bruce Wielicki, NASA Langley Research Center

accessible EOS data, with research results available through EOSDIS to enhance broad participation by the science community at large.

The EOS Program will help us understand the Earth's environment and will provide policymakers with critical insight into its expected evolution. Unfortunately, EOS cannot pursue some important scientific questions that it would have addressed under the original budget. Several instruments had to be de-selected, and others were left without EOS funding and designated for "flights of opportunity." A reduction in the EOS budget—from $17 to $7 billion over the life of the mission—mandated such economies. For this reason, the EOS research emphasis shrank from global change to global *climate* change. Even taking into account the diminished, yet more focused, science return, the EOS Program will generate an unprecedented amount of Earth science data.

Obviously, a wealth of raw data is not useful without the expertise to manipulate the data streams into usable geophysical products. The instrument and IDS teams provide this expertise, and will share their knowledge via EOSDIS with Earth scientists around the world. Moreover, the EOS Program has already begun to address global change by extending and improving existing data sets. Previous satellite and other data sets are being re-assessed in order to produce a longer baseline from which researchers can determine the rate of global change.

Archived data sets are being transferred to more stable media and are being reprocessed in standard formats to ensure compatibility with future observations. Critical global satellite data that reside in Federal archives have already been identified by the EOS Program as "Pathfinder" projects: Advanced Very High-Resolution Radiometer (AVHRR), TIROS Operational Vertical Sounder (TOVS), Geostationary Operational Environmental Satellite (GOES), Special Sensor Microwave/Imager (SSM/I), Scanning Multispectral Microwave Radiometer (SMMR), and Land Remote-Sensing Satellite (Landsat) data sets. The EOS Program seeks to make research-quality global change data sets available to the Earth science community as soon as possible. These Pathfinders follow the successful reprocessing of Nimbus-7 Total Ozone Mapping Spectrometer (TOMS) data to establish ozone trends and to track volcanic sulfur dioxide concentrations in the atmosphere, and ocean phytoplankton biomass time series from the Coastal Zone Color Scanner (CZCS) in anticipation of the launch of the Sea-Viewing Wide Field Sensor (SeaWiFS) and the EOS satellites.

By enhancing the utility of present holdings, the EOS Program assists global change researchers now. Chapters 2 through 8 discuss global change science issues in light of present data holdings and the observations to be secured by the EOS-era remote-sensing instrumentation.

SUGGESTED READING

Asrar, G. and D.J. Dokken (eds.), 1993: *EOS Reference Handbook*, National Aeronautics and Space Administration, Washington, D.C., NP-202, 145 pp.

Dozier, J., 1994: Planned EOS observations of the land, ocean, and atmosphere. *Atmospheric Research*, 31, pp. 329-357.

Earth System Sciences Committee, 1988: *Earth System Science: A Closer View*. NASA Advisory Council, University Corporation for Atmospheric Research, Boulder, Colorado, 208 pp.

Houghton, J.T., B.A. Callander, and S.K. Varney (eds.), 1992: *1992 IPCC Supplement: Scientific Assessment of Climate Change*. Submission from Working Group I. Cambridge University Press, Cambridge, 24 pp.

National Research Council, 1991: *Opportunities in the Hydrologic Sciences*. Committee on Opportunities in the Hydrologic Sciences, Water Science and Technology Board, National Academy Press, Washington, D.C.

CHAPTER 2

Clouds, Radiation, Water Vapor, and Precipitation

The range of surface temperatures and pressures on Earth is such that water is plentiful in its life-supporting liquid state and yet moves freely and vigorously to its vapor and solid states as well. The more we learn about our desiccated, and apparently barren, neighboring planets, the more we wonder if our good fortune is not a result as well as the cause of life on the Earth.

Opportunities in the Hydrologic Sciences, 1991

CURRENT UNDERSTANDING

The highest priority science and policy issues confronting EOS are those concerning potential changes in the Earth's water and energy cycles. The science community now agrees that the Earth's climate will undergo changes in response to the radiative forcing induced by the increase in trace gases, by changes in tropospheric aerosols, and by solar variability. These changes may profoundly affect atmospheric temperature, humidity, circulation, cloud cover distribution, and especially precipitation patterns. Such changes are certain to have significant societal impact. Water is central to the functioning of human society in ways that only become apparent when it is unusually scarce or abundant. In the U.S. alone, half of the $100 billion agricultural industry depends on irrigation, while 90 percent of U.S. electricity generation involves water in some form. Climate models already suggest a future tendency toward increased aridity and drought, particularly in the interiors of mid-latitude continents.

The full impact of greenhouse gases and aerosol forcing depends on poorly understood feedback between atmospheric changes and radiative forcing terms. For example, increased atmospheric moisture caused by greater evaporation at the Earth's surface is a near certainty in a warmer climate. At present, we are uncertain about the resulting vertical redistribution of this evaporation, which is essential in understanding the magnitude of water vapor feedback.

The monitoring of clouds, water vapor, aerosols, and the radiation budget of the planet is therefore a key component of a space-based Earth-observing system. The Earth's radiation budget is especially important because it serves as a tracer of climate change and a forcing function. Understanding the disposition of radiative energy within the climate system is essential, because the solar and longwave fluxes at the top of the atmosphere and the Earth's surface are primary components of the Earth's radiation budget. The energy supplied to the Earth by the Sun (i.e., the total solar irradiance), atmospheric chemical composition, and the distribution of oceans and land masses combine to determine the radiative balance, hence the climate of the biosphere.

Because the spatial variation of radiative heating drives the general circulation of the atmosphere-ocean system, changes in this heating will cause changes in circulation patterns. These will alter the hydrologic cycle and the distribution of precipitation over the surface of the Earth. Two primary areas require intensive study to better understand the Earth's energy and water cycles: Clouds and atmospheric circulation.

Clouds

Clouds are the dominant modulator of the flow of radiation between the Sun, the Earth's surface, and the flow back to outer space. A fundamental question is "Do clouds cool or heat the current climate?" Low, optically thick clouds (e.g., those found off the western coast of South America or Africa) or the low clouds in mid-latitude storm systems strongly reflect sunlight. These clouds' temperatures are nearly equal to the surface temperatures, so the emitted longwave flux is about the same. High, cold, optically thin cirrus clouds reduce the emission of longwave flux to space, and thereby warm the surface and lower atmosphere. Until the Earth Radiation Budget Experiment (ERBE), scientists did not know whether the cooling effect dominated the heating effect. ERBE measurements demonstrated that clouds usually cool the current global climate.

Clouds respond to changes in their environment. As we change the Earth's climate, cloud properties and their influence on the radiation budget will also change. A critical question is "How sensitive are cloud formation and dissipation processes and cloud radiative properties to changing radiative perturbations or climate forcing?" Currently, cloud parameterizations in global models introduce a three-fold uncertainty in the predicted magnitude of global surface temperature change resulting from more abundant greenhouse gases.

While clouds dominate the modulation of the energetics of the climate system, the presence of clouds and cloud optical properties themselves depend on the climate. Even if we know the optical properties and spatial distribution of the clouds, there is a moderate uncertainty in their radiative effect on the energy budget. There are still fundamental problems connected with the calculation of the radiation fields when there is spatial inhomogeneity, such as horizontally broken or vertically overlapping clouds. In addition, accurate calculations often require information about the surface that is not readily available, such as the angular dependence of the surface reflectance and its dependence on winds, vegetation, and numerous other factors.

The development of increasingly sophisticated models for the study of climate forcing has resulted in substantially more sensitivity to boundary conditions. The ability of climate models to accurately predict global climate change demands that the physical correctness of their assumptions and algorithms be verified against the empirical record of climate variability. For example, the solar irradiance database has identified time scales for the solar cycle based on a direct correlation of luminosity and solar activity. Subtle trends in total irradiance, as little as ±0.5 percent per century, could eventually produce the extreme range of climates known to have existed in the past—from warm periods without permanent ice to the great Ice Ages. Variation in solar irradiance, which could ameliorate or exacerbate global warming, needs to be studied in tandem with cloud formation and dissipation to determine the net effect.

We understand the dependence of cloud parameters on the variables of the climate system only in isolated areas under limited conditions. Observational efforts to increase our understanding, including the campaigns of the First ISCCP Regional Experiment (FIRE), have demonstrated the complexity of cloud atmospheric physics through simultaneous measurements of cloud properties and radiation fields. Unfortunately, cloud optical properties usually depend on the availability of water and the presence of vertical updrafts. These two variables are among the most difficult to measure either *in situ* or remotely, and they are difficult to model because the water content of a parcel is sensitive to its temperature and because the vertical velocity is typically part of a complex and chaotic turbulence field.

At the local scale, suitable long-term observational sites, such as those of the Atmospheric Radiation Measurement (ARM) Program of the Department of Energy, provide scientists with cloud and radiation data. However, only space-based observations, with the stable calibration capability and simultaneity of the EOS instrument and satellite combinations, will provide the global perspective needed to respond with certainty to questions about how clouds respond to and foster climate change.

Atmospheric Circulation

Atmospheric circulation conveys moisture and heat. Water vapor evaporates from the ocean and land surfaces, then is transported to other parts of the atmosphere where condensation occurs, clouds form, and precipitation ensues. There is a net outflow of atmospheric moisture from the tropics, where sea-surface temperatures are high and the warm atmosphere can hold large amounts of water vapor, to the higher latitudes, where condensation and precipitation remove the moisture. A warmer climate will affect the hydrologic cycle because the amount of moisture that circulates through the atmosphere will be greater. However, current models give inconsistent results about the future distribution of precipitation.

Water vapor concentration depends on temperature, which determines the total amount of water that the atmosphere can hold without saturation. Hence, water vapor amounts decrease from equator to pole and with increasing altitude. Superimposed on this general behavior are smaller variations of water vapor amounts that determine the formation and properties of clouds and rainfall. Generally, water amounts are less over the continents than over the oceans, and the upper atmosphere is drier than the near-surface atmosphere. Researchers do not yet understand all the processes that transport water accurately enough to pre-dict how variations in water vapor will change with temperature. Simultaneous imaging of clouds and water vapor from satellites can provide the data necessary to better understand the interactions among cloud processes and the large-scale flow and distribution of water vapor.

The oceans and the land are as important as the atmosphere in determining the global energy and hydrologic cycles. Ocean-atmosphere interactions and associated ocean circulation patterns receive attention in Chapter 3. Over the land, more needs to be known about how energy and water fluxes link with the state of land and its vegetation. Many of the uncertainties reflect the lack of synergistic measurements of soils, soil moisture, vegetation morphology and physiology, snow depth, and the state of the overlying atmosphere. Direct measurements of evaporation and transpi-ration are available only at local sites; regional and global parameterizations of these critical fluxes are currently inadequate. Global climate models incorporate the interactions of the land surface and atmosphere in only a primitive way.

KEY QUESTIONS

- How will clouds influence the Earth's radiation budget, and modulate or enhance tropospheric warming caused by increasing concentrations of greenhouse gases, especially water vapor?

- How will tropospheric warming from increasing greenhouse gases affect atmospheric circulation patterns and regional precipitation patterns?
- How do increasing concentrations of atmospheric trace gases affect the global and regional balances of precipitation, evaporation and transpiration, and runoff?
- What are the feedback and interactions between the atmosphere, oceans, and land-surface hydrology?
- How do surface fluxes and large-scale atmospheric motions transport moisture and partition energy among their various sinks and sources, and what are the effects of these transport processes on regional climate fluctuations?
- Does the variation of solar irradiance during the solar cycle modulate greenhouse forcing, and is there a systematic long-term variation in solar irradiance leading to different climate regimes like the "Little Ice Age?"

SCIENCE STRATEGY

During the 1990s, coordinated experiments such as ISCCP, FIRE, ARM, and ERBE will improve knowledge of cloud behavior, while the Global Energy and Water Cycle Experiment (GEWEX) will address large-scale hydrologic issues.

The joint U.S./Japanese Tropical Rainfall Measuring Mission (TRMM) mission, to be launched in 1997, will mark the beginning of improved measurements of precipitation over regions where current data are inadequate. In tandem with EOS instruments, TRMM will acquire crucial information on surface fluxes of water and the transport of water vapor through the atmosphere. Before EOS, TRMM will provide unprecedented simultaneous measurements of latent heating and radiation. The role of aerosols in moderating greenhouse forcing will be investigated using data from the French Polarization and Directionality of Earth's Reflectance (POLDER) instrument on the Advanced Earth Observing System (ADEOS) satellites sponsored by Japan (see Chapter 9).

The most important EOS instruments for information on water and energy cycles include ACRIM, AIRS/AMSU/MHS, CERES, EOSP, GLAS, MIMR, MISR, MLS, and MODIS. EOSP, MISR, MODIS, and SAGE III will provide new information on aerosol formation and distribution. The combination of ACRIM, CERES, and MODIS will greatly improve estimates of the large-scale and low-frequency variability of net incoming solar radiation and net outgoing longwave radiation and their connection to cloud structure and coverage. MISR will globally measure the shortwave reflection and scattering properties of clouds and the surface at scales of a few kilometers, and determine how absorption of solar radiation varies with the detailed characteristics of clouds and land vegetation.

MLS—already in orbit aboard the Upper Atmosphere Research Satellite (UARS)—obtains global maps of upper tropospheric water vapor on a daily basis, even in the presence of volcanic aerosols and high cirrus clouds in the equatorial regions.

Nimbus-7 (1978-1993) began the solar irradiance database, and the more precise Solar Maximum Mission (SMM)/ACRIM I experiment (1980-1989) refined it. The UARS/ACRIM II experiment is currently continuing the data record. This instrument should function through 1995, providing sufficient overlap with the next irradiance-monitoring experiment—the Solar and Heliospheric Observatory (SOHO), to be launched in mid-1995. ACRIM is next scheduled for deployment in 2003, onboard EOS-CHEM1.

In tandem with EOSP and SAGE III, MISR and MODIS will explore the role of aerosols in radiation balance. Understanding cloud formation processes and inter-action will be advanced by the AIRS/AMSU/MHS, GLAS, and MODIS instruments. In addition, MIMR and MODIS sensors together will provide global mapping of snow cover in all weather conditions.

In addition to the teams of scientists supporting each of the instruments listed above, several interdisciplinary science studies focus on clouds, radiation, water vapor, and precipitation, including:

- *Global Water Cycle: Extension Across the Earth Sciences*, Eric Barron
- *NCAR Project to Interface Modeling on Global and Regional Scales with EOS Observations*, Robert Dickinson
- *Interannual Variability of the Global Carbon, Energy, and Hydrologic Cycles*, James Hansen
- *Climate Processes over the Oceans*, Dennis Hartmann
- *The Hydrologic Cycle and Climatic Processes in Arid and Semi-Arid Lands*, Yann Kerr and Soroosh Sorooshian
- *Global Hydrologic Processes and Climate*, William Lau
- *The Processing, Evaluation, and Impact on Numerical Weather Prediction of AIRS, AMSU, and MODIS Data in the Tropics and Southern Hemisphere*, John LeMarshall
- *The Role of Air-Sea Exchanges and Ocean Circulation in Climate Variability*, W. Timothy Liu
- *An Interdisciplinary Investigation of Clouds and the Earth's Radiant Energy System: Analysis*, Bruce Wielicki.

An overview of a typical EOS study illuminates how teams of researchers attack global change issues. Dr. Wielicki's investigation uses existing data to provide the

EOS Program with a consistent benchmark database of accurately known fields of radiation and cloud properties. EOS will extend this database by discerning radiative fluxes at the top of the Earth's atmosphere, at the Earth's surface, and as flux divergence within the atmosphere. Cloud properties will include measured spatial coverage, cloud altitude, shortwave and longwave optical depths, cloud particle size, and condensed water density. Radiative transfer models and data analysis algorithms exploiting AVHRR, High-Resolution Infrared Sounder (HIRS), and ERBE data sets—along with the field measurements of clouds and radiation collected and analyzed during the FIRE campaign—will populate the benchmark database, and help focus interdisciplinary research objectives in anticipation of EOS-era observations.

The *EOS Reference Handbook* (Asrar and Dokken, 1993) contains synopses of the other investigations listed above. For deeper coverage of the science currently pursued by these interdisciplinary science teams, consult the articles cited at the close of this chapter.

EXPECTED RESULTS

The EOS Program seeks to advance knowledge about hydrologic and energy systems by exploiting both pre-EOS and EOS-era observations, and associated modeling efforts to interpret these measurements. Examples of key parameters to be honed for input into these global change models follow:

1) **Changes in the long-term baseline of cloud radiative forcing**—The ERBE determination of cloud radiative forcing provides a baseline for detecting future change. Although the data record is short, the documented ERBE data set is accepted by the community, so comparisons with EOS-era observations will receive priority. One set of CERES data products specifically will use the validated ERBE algorithms to provide cloud forcing measurements nearly identical to the ERBE data set. The only differences will arise because of changes in the detectors and their calibrations. Changes in cloud forcing will be detected within a few percent. ACRIM will extend the high-precision solar irradiance database, facilitating analysis of its climate implications on multi-decadal time scales.

2) **Cloud feedback mechanisms**—Cloud feedback involves the derivation of cloud forcing that specific changes in the climate system drive. Such derivatives can be investigated by analyzing the response of clouds to the seasonal cycle in solar input and thermal response; by studying the differential response of clouds to various underlying conditions, such as ocean circulation and vegetation regime; by examining the response of clouds to sea ice

and snow advance and retreat; and by checking the response of clouds to transient events, such as large emissions of sulfate aerosols from continents.

3) **The role of clouds in modulating the surface radiation budget of vegetated regions**—Just as clouds modulate the radiation budget at the top of the atmosphere, they also modulate the radiation budget of the Earth's surface. Recent advances in data analysis techniques have led to methods that accurately deduce the net flux of shortwave energy at the Earth's surface from radiation measurements at the top of the atmosphere. Such modulation strongly influences the overall energy budget and affects the contrast between north- and south-facing slopes. In regions of frequent overcast, there may be little north-south contrast, and the vegetation of the region may be homogeneous. On the other hand, in deserts, there is a strong contrast between the radiation balance of northern and southern slopes. The vegetation of regions dominated by clear skies will clearly show this difference. In the EOS era, the satellite-derived surface radiation budget will reveal the influence of such regimes on both flora and fauna of the Earth's surface. The modulation of the solar radiation field at the Earth's surface by clouds strongly influences the rate of photosynthesis, thus the carbon budget. Current models do not usually include such cloud modulation; it is likely to add significant insight into the proposed mechanisms of climate change.

4) **The interaction between clouds and the energy budget of the oceans**— Clouds modulate the influx of solar energy into the surface layers of the oceans. At present, *in situ* measurements of that energy flux and estimates of the heat transport required to balance the satellite measurements of radiation input and losses differ by substantial amounts. The EOS measurements of radiation at the top of the atmosphere and at the ocean surface will improve understanding of the source of this discrepancy, thus the ability to model heat and water transport. EOS will provide information on the extent of cloudiness and distinguish cloud properties over regions not readily accessible by aircraft or surface observations. Because clouds both move and vary, only satellite measurements can determine whether simple cloud motions or more complex changes in cloud properties cause the observed variability.

5) **Types of clouds and meteorological conditions most important in determining cloud forcing and feedback**—Clouds are semi-discrete "objects" that form and dissipate on short time scales, yet also are participants in large-scale meteorological flows by changing atmospheric heating rates and surface thermodynamic conditions. EOS research into cloud-radiation interactions will generate new categories for understanding the conditions under which clouds dominate the flow fields, and the conditions under which the clouds are passive indicators of meteorological flows. Because the morphology and types of clouds indicate the nature of

the underlying physical phenomena, satellite-derived cloud types can also play a role in improving meteorological models of cloud formation, maintenance, and dissipation. These improvements, in turn, will lead to improvements in the parameterization of clouds in models.

6) **Continental-scale water and energy exchanges**—The distribution of the global radiation budget over the Earth's surface has a strong influence on the transfer of latent and sensible heat. In addition, the solar radiation input to the vegetation on land and phytoplankton in the ocean surface layer plays a key role in the transfer of water and carbon from the Earth's surface to the atmosphere. The near-simultaneity of instruments in the EOS satellite series will help identify causal relationships in the hydrologic cycle. Time variations in the surface energy budget can be examined, and can be compared with regions in similar latitudes to help isolate the influence of clouds and water vapor on the hydrologic cycle.

7) **Continental and regional surface hydrologic models with explicit treatment of precipitation, runoff, and snow and ice dynamics over land areas**—EOS data products will substantially improve knowledge of cloud water and ice content, and cloud vertical structure within the atmosphere. The derivation of radiation fluxes at the top of the atmosphere, at the surface, and at cloud tops can determine how the radiation field varies with time over a particular geographic region. With the aid of surface reflectance from high-resolution spectral channels and of the angular characteristics of the surface, researchers should be able to improve the detection of precipitation events, including snowfall and perhaps rainfall (with less precision). Vegetation change after such precipitation events may improve understanding of how the surface and radiation characteristics modulate the overall hydrology of various regions of the Earth.

8) **High-resolution, comprehensive Earth system models that incorporate atmosphere-ocean interactions along with realistic treatments of clouds and the hydrologic cycle**—EOS data will provide substantial, quantitative improvements in understanding the time and space variations of both clouds and the underlying surface and circulation patterns. Observations of both short- and long-term variations are critical in developing adequate parameterizations of the phenomena involved. Modelers need to know whether the clouds forced radiation variations before other phenomena changed, or whether the clouds simply responded after other changes had taken place.

SUGGESTED READING

Albrecht, B.A., 1993: Effects of precipitation on the thermodynamic structure of the trade wind boundary layer. *Journal of Geophysical Research*, 98, pp. 7327-7337.

Bennett, A.F., L.M. Leslie, C.R. Hagelberg, and P.E. Powers, 1993: Tropical cyclone prediction using a barotropic model initialized by a generalized inverse method. *Monthly Weather Review*, 121, pp. 1715-1729.

Cess, R.D., E.F. Harrison, P. Minnis, B.R. Barkstrom, V. Ramanathan, and T.Y. Kwon, 1992: Interpretation of seasonal cloud-climate interactions using Earth Radiation Budget Experiment data. *Journal of Geophysical Research*, 97, pp. 7613-7617.

Fu, R., A.D. Del Genio, W.B. Rossow, and W.T. Liu, 1992: Cirrus-cloud thermostat for tropical sea-surface temperatures tested using satellite data. *Nature*, 358, pp. 394-397.

Klein, S.A. and D.L. Hartmann, 1993: The seasonal cycle of low stratiform clouds. *Journal of Climate*, 6, pp. 1587-1606.

Klein, S.A. and D.L. Hartmann, 1993: Spurious changes in the ISCCP data set. *Geophysical Research Letters*, 20, pp. 455-458.

Lau, W.K.-M. and L. Peng, 1992: Dynamics of atmospheric teleconnection during northern summer. *Journal of the Atmospheric Sciences*, 5, pp. 140-158.

Liu, W.T., W. Tang, and F.J. Wentz, 1992: Precipitable water and surface humidity over global oceans from Special Sensor Microwave Imager and European Center for Medium Range Weather Forecasts. *Journal of Geophysical Research*, 97, pp. 2251-2264.

Mapes, B.E. and R.A. Houze, Jr., 1993: Cloud clusters and superclusters over the oceanic warm pool. *Monthly Weather Review*, 121, pp. 1398-1415.

Rossow, W.B., 1993: Comparison of ISCCP and other cloud amounts. *Journal of Climate*, 6, pp. 2394-2418.

Sohn, B.-J. and F.R. Robertson, 1993: Intercomparison of observed cloud radiative forcing: A zonal and global perspective. *Bulletin of the American Meteorological Society*, 74, pp. 997-1006.

Spencer, R.W. and J.R. Christy, 1993: Precision lower stratospheric temperature monitoring with the MSU: Technique, validation, and results 1979-1991. *Journal of Climate*, 6, pp. 1194-1204.

Sud, Y.C., W.C. Chao, and G.K. Walker, 1992: Role of cumulus convection in maintaining atmospheric circulation and rainfall–a GCM simulation study. *Monthly Weather Review*, 120, pp. 594-611.

Wang, W.-C., G.-Y. Shi, and J.T. Kiehl, 1991: Incorporation of the thermal radiative effect of CH_4, N_2O, CF_2Cl_2, and CFC_{13} into the National Center for Atmospheric Research community climate model. *Journal of Geophysical Research*, 96, pp. 9097-9103.

Welch, R.N., S.K. Sengupta, A.K. Goroch, P. Rabindra, N. Rangaraj, and N.S. Navar, 1992: Polar cloud and surface classification using AVHRR imagery: An intercomparison of methods. *Journal of Applied Meteorology*, 31, pp. 405-420.

Wielicki, B.A. and L. Parker, 1992: On the determination of cloud cover from satellite sensors: The effect of sensor spatial resolution. *Journal of Geophysical Research*, 97, pp. 12799-12824.

CHAPTER 3

Oceans: Circulation, Productivity, and Air-Sea Exchange

Today, realistic models of the ocean are impossible to construct owing both to uncertainty over the governing physics, chemistry, and biology, and an inadequate ability to prescribe the present state of the system. Global, continuous observations of the ocean are required.

The Future of Spaceborne Altimetry: Oceans and Climate Change, 1992

CURRENT UNDERSTANDING

The oceans are an enormous reservoir of heat and carbon, transporting significant mechanical and chemical energy and driving long-term changes in climate. Researchers acknowledge that the oceans are a dominant element of climate change, but they are just beginning to fathom some basic principles underlying ocean dynamics and productivity.

Simply stated, the global system of ocean currents can be compared to a giant conveyor belt that moves heat, salt, and chemicals around the planet. A better understanding of how this belt functions needs to be developed if researchers are to build accurate general circulation models that couple the ocean and atmosphere. For instance, the oceans carry from one-third to one-half of the heat from equator to pole, a flux that renders the mid-latitudes of the Earth warm enough to be habitable. Ocean-atmosphere climate models try to analyze the global effects of changing ocean variables; for example, an increase in precipitation over the ocean would reduce convective mixing and bottom-water formation. However, without precise parameterizations of the involved processes, global change scenarios are suspect.

Oceanography used to be limited to ship and buoy operations, which yielded a patchy, regionalized perspective. Remote sensing from space now provides a global, repeatable view, through existing missions such as the Ocean Topography Experiment (TOPEX/Poseidon) and the upcoming constellation of EOS satellites. The EOS Program has developed an interdisciplinary framework to exploit and carefully calibrate historical observations (e.g., CZCS) with those expected soon (e.g., SeaWiFS). This framework efficiently integrates focused field studies; advanced methods for product generation, archive, and distribution; and observations from space.

The world ocean serves as a huge reservoir of elements and chemical species to which the climate system is sensitive. In particular, the ocean is a dynamic storehouse for carbon, holding approximately 60 times as much carbon (37,000 gigatons as dissolved inorganic carbon, nearly 1,000 gigatons in dissolved organic forms, and about 30 gigatons as particulate carbon) as the atmosphere (650 gigatons in carbon dioxide). The dissolved organic carbon alone equals the total vegetative carbon reservoir on land, and while the total amount of phytoplankton (particulate carbon) in the ocean at any one time is a small fraction of the biomass of terrestrial plants, we estimate that through photosynthesis they fix about as much carbon annually as land plants do. The incorporation of large amounts of inorganic carbon by marine plants is important in the cycling of carbon and other elements in the ocean, but the controlling processes (or those influenced by such uptake) are not well-understood. The spatial and temporal variation in carbon cycling remains a mystery. Formation and destruction of dissolved organic matter—hence its concentration in particular regions—depend on inputs from rivers, bottom sediments, and plant species, which in turn depend on terrestrial and oceanic processes. The roles of phytoplankton and dissolved organic carbon in balancing the global carbon budget is a principal question that EOS investigators seek to answer, since oceanic carbon uptake is a possible mechanism to delay the greenhouse effect.

Oceans contain 98 percent of the available water in the world, stored in both its fluid and solid states (i.e., sea ice). About 78 percent of global precipitation and 86 percent of evaporation occurs over the oceans (Baumgartner and Reichel, 1975). The small net evaporation over the oceans supplies the net precipitation over continents,with regional patterns of evaporation and precipitation controlling upper ocean density. Fluxes of energy, momentum, and freshwater between the atmosphere and the ocean surface affect circulation and the availability of both water vapor and energy to the atmosphere. Ocean circulation patterns transport oceanic heat and interact with atmospheric circulation to cause changes in weather patterns and continental precipitation, thereby strongly influencing regional economic activity.

Phenomena at the ocean surface and the surface layer of the atmosphere tightly control transfer of carbon and many other chemicals across the air-sea interface.

Exchange of carbon dioxide across this interface depends on the difference of the partial pressures of carbon dioxide in air and water, wind stress, and how fast the gas can penetrate the sea surface. The penetration rate depends in part on surface organic slicks derived from phytoplankton blooms, rivers, and other coastal environs. The difference of partial pressure and wind stress at the sea surface controls the mechanics of exchange, and plants play an active role. Phytoplankton help regulate the partial pressure of carbon dioxide in the water and support a diverse ecological structure that profoundly affects the health of the planet. The oceans support robust ecosystems that provide recreational and nutritional benefits to humankind, and projections of the socio-economic impact of global change must include changes in these ecosystems. Oceanic processes control global change on many levels—including the human dimension, which ultimately drives policy formulation.

A good conceptual understanding of oceanic processes now exists, as do good estimates of the larger reservoirs of elements, heat, and momentum. Unfortunately, measurement techniques and strategies before modern missions left lingering uncertainties, especially in the size of the fluxes of elements, heat, and momentum. Researchers must address the fate of anthropogenically produced carbon and anomalies in the amount of heat transported north or south if they hope to determine the oceans' effect on weather patterns. To better understand the role of the oceans in global change, scientists need data about physical, chemical, and biological processes influencing the climate-sensitive species in the upper ocean, and about the processes that control ocean-atmosphere interactions.

KEY QUESTIONS

- What are the large-scale transports of heat and water within the upper ocean, and how do they control air-sea exchange and vertical ocean circulation?
- What dynamical balances control upper ocean circulation, and how will the circulation change in response to changing surface fluxes?
- What are the oceanic biogeochemical cycles for climate-sensitive (chemical) species, and how sensitive are these cycles to changes in physical climate?
- What are the mechanisms for, and present magnitudes of, air-sea exchanges of climate-sensitive chemical species?
- What are the rate, geographical distribution, and controlling factors for oceanic primary production and new production?
- How does ecosystem structure control the rate of organic matter transport to deep oceans?
- How are oceanic ecosystems changing in structure and species makeup?

SCIENCE STRATEGY

Enormously valuable data sets about ocean surface properties are becoming available from satellites. Satellites do not provide data on processes in the ocean interior, but they provide critical observations for understanding the interaction of the ocean with the atmosphere, thus the whole climate system. Satellite remote sensing of the oceans has a varied, brief history. The Defense Meteorological Satellite Program (DMSP) SSM/I sensor has collected observations of integrated atmospheric liquid water, water vapor, and surface wind speed since 1987, and these are used to estimate monthly mean air-sea fluxes of heat and water. SSM/I observations of sea ice—along with synthetic aperture radars (SARs) on the European Remote-Sensing Satellite-1 (ERS-1) and Japan's Earth Resources Satellite-1 (JERS-1)—are input for models of bottom-water formation and heat, mass, and property fluxes in the high-latitude oceans. Since 1992, TOPEX/Poseidon has provided a wealth of data about the circulation of the upper ocean. The NOAA AVHRR provides high-resolution sea-surface temperature in cloud-free regions. Satellite data, coupled with field observations from studies such as the World Ocean Circulation Experiment (WOCE) and Joint Global Ocean Flux Study (JGOFS), enable significant advances in understanding the role of the ocean in the Earth system.

Starting in early 1995, SeaWiFS will routinely provide ocean color measurements, from which biomass and biological productivity can be estimated. The U.S./Canadian Radar Satellite (Radarsat) will augment the international SAR capability in 1995. In addition, the NASA Scatterometer (NSCAT) and Ocean Color and Temperature Scanner (OCTS) will fly on the Japanese ADEOS mission in 1996, to measure wind stress on the ocean surface and to provide additional measurements of ocean color and productivity, respectively.

In 1998 and beyond, MODIS sensors flying as part of the EOS Program will continue the SeaWiFS ocean color time series and improve the accuracy and resolution of sea-surface temperature measurements. Subsequently, more complete spectral measurements, by sensors such as the Global Imager (GLI) to fly on ADEOS II in 1999, will allow separation of phytoplankton pigment into its critical components, significantly improving understanding of ocean biogeochemical cycles.

General circulation models are incorporating estimates of global air-sea exchange of carbon, using techniques from turbulent parameterization to four-dimensional assimilation of observations. The exchanges of carbon or fluxes depend on wind speed (measured by scatterometers or microwave radiometers) and the sea-air difference in the partial pressure of carbon dioxide. In the atmosphere, a blend of general circulation model output and climatological observations must provide values for the partial pressures of carbon dioxide. Ocean surface carbon dioxide partial pressure depends on wind stress (transport), ocean color pigment (organic sources), and sea-surface pressure (storage capacity). Parameterizations based on data from instruments such as NSCAT, SeaWiFS, and AVHRR are currently being explored. ADEOS II/SeaWinds will continue the ocean vector wind data set, allowing improved estimation of critical air-sea fluxes of heat, momentum, and chemical species.

AIRS/AMSU/MHS will provide accurate sea-surface temperatures, as well as boundary-layer temperature and moisture needed for monitoring air-sea energy exchange. MIMR will provide sea-ice concentration and atmospheric water content at a finer spatial resolution than SSM/I, and its 6 GHz channel will allow estimation of sea-surface temperature under cloud cover. Continuous time series of altimetric measurements will allow the study of critical components of low-frequency variability of basin-scale ocean circulation.

In addition to the scientists supporting each of the instruments listed above, several interdisciplinary research teams focus on the role of the ocean in the climate system:

- *Coupled Atmosphere-Ocean Processes and Primary Production in the Southern Ocean*, Mark Abbott
- *Biogeochemical Fluxes at the Ocean/Atmosphere Interface*, Peter Brewer
- *Interdisciplinary Studies of the Relationships between Climate, Ocean Circulation, Biological Processes, and Renewable Marine Resources*, J. Stuart Godfrey
- *Climate Processes Over the Oceans*, Dennis Hartmann
- *Global Hydrologic Processes and Climate*, William Lau
- *The Role of Air-Sea Exchanges and Ocean Circulation in Climate Variability*, W. Timothy Liu
- *Investigation of the Atmosphere-Ocean-Land System Related to Climatic Processes*, Masato Murakami
- *Polar Exchange at the Sea Surface (POLES): The Interaction of Oceans, Ice, and Atmosphere*, Drew Rothrock
- *Middle and High Latitudes Oceanic Variability Study*, Meric Srokosz
- *Earth System Dynamics: The Determination and Interpretation of the Global Angular Momentum Budget Using EOS*, Byron Tapley.

A brief synopsis of a typical interdisciplinary science investigation—research by W. Timothy Liu's team—illustrates how the EOS Program contributes to the understanding of oceanic processes. This team seeks to improve methods for monitoring ocean-atmosphere responses to exchanges in fluxes of momentum, energy, and moisture. They have shown that variations of evaporation that accompany the interannual El Niño-Southern Oscillation (ENSO) are as large as cloud effects in determining the surface energy balance of the tropical ocean. Thus, interannual variations of sea-surface temperature are not controlled simply by a cirrus cloud "thermostat." This finding proves critical as researchers attempt to develop general circulation models that accurately depict the storage and release of latent heat, a principal mechanism that influences atmospheric circulation and that globally redistributes both water and energy. They are now focusing on development of eddy-resolving general circulation models (including thermodynamics) to obtain a four-dimensional description of such storage and transport and of greenhouse gases in the ocean. This team currently uses data from the TOPEX/Poseidon and ERS-1 satellites, which will be continued by the ADEOS missions and the EOS series well into the next century.

The *EOS Reference Handbook* (Asrar and Dokken, 1993) contains synopses of the other investigations listed above. For deeper coverage of the science currently pursued by these interdisciplinary science teams, consult the articles cited at the close of this chapter.

EXPECTED RESULTS

The next decade will see many advances in understanding the role of oceans in modulating the Earth system. The following list encapsulates some of the more striking accomplishments expected during the EOS time frame:

1) **Significant improvements in the quality, coverage, and availability of ocean data sets on a global basis**—The pre-EOS era will witness advances in single oceanographic disciplines (e.g., biology, chemistry, and physics), while the EOS era will focus on the complex, non-linear interaction of these systems.

2) **Significant improvement in coupled atmosphere-ocean models**— Toward the end of the 1990s, scientists will develop coupled atmosphere-ocean models capable of assimilating complex data fields. Because the links between these two systems are so poorly understood now, these coupled models may well be the most important advance in global change research. In the EOS era, predictive models will become operational through a combination of experience in working with earlier models, the continued evolution in understanding the relevant processes, and the beginning of a consistent time series of observational data accessible via EOSDIS. The models in the EOS era will contain more processes explicitly, and the parameterizations that remain will be better understood.

Continued effort will be placed on developing fully predictive models of ocean biogeochemistry.

3) **Detailed determinations of the accuracy of scatterometer wind measurements**—This task is critical, because eventual use of remotely sensed data to force and test advanced models requires quantitative estimates of data accuracy to allow proper assimilation and to ensure proper interpretation of results. This involves the development and testing of techniques for extrapolating and interpolating scatterometer data to regular space-time grids and establishment of quantitative accuracies of products. Researchers expect to generate several basin-scale and regional ocean wind products for the Atlantic, North Pacific, Tropical Oceans, and Southern Ocean. Key areas of near-term study include large-scale tropical air-sea interaction modeling, the Southern Ocean, and biophysical modeling of the Gulf Stream region of the Atlantic.

4) **Improved estimation of ocean-atmosphere exchanges**—At present, estimation of ocean-atmosphere exchanges in momentum, shortwave radiation, and latent heat flux approach an accuracy sufficient to delineate seasonal cycles over most of the ocean and interannual variations in the tropical oceans. Satellite derivations of monthly mean precipitation patterns over the tropical oceans are now possible. Improvement in quantitative estimation of freshwater exchange is expected from the instruments aboard TRMM, and AMSU and MIMR aboard the EOS-PM series. Longwave radiation and sensible heat flux data will be enhanced with the launch of AIRS, also aboard the EOS-PM series. With sensor and algorithm improvements in the future, absolute accuracy in surface heat flux sufficient to evaluate the mean meridional ocean heat transport is anticipated. Assimilation of data from ocean-observing satellites into operational numerical weather prediction models soon also will improve the presently weak estimation of surface fluxes.

SUGGESTED READING

Baumgartner, A. and E. Reichel, 1975: *The World Water Balance: Mean Annual Global, Continental, and Maritime Precipitation, Evaporation, and Runoff.* Elsevier Scientific Publishing Company, Amsterdam, 179 pp.

Bennett, A.F. and J.R. Baugh, 1992: A parallel algorithm for variational assimilation in oceanography and meteorology. *Journal of Atmospheric and Oceanic Technology,* 9, pp.426-433.

Chelton, D.B. and M.G. Schlax, 1993: Spectral characteristics of time-dependent orbit errors in altimeter height measurements. *Journal of Geophysical Research*, 98C, pp. 12579-12600.

Endoh, M., T. Tokioka, and T. Nagai, 1991: Tropical Pacific sea-surface temperature variations in a coupled atmospheric-ocean general circulation model. *Journal of Marine Systems*, 1, pp. 293-298.

Freilich, M.H. and R.S. Dunbar, 1993: Derivation of satellite wind model functions using operational surface wind analyses: An altimeter example. *Journal of Geophysical Research*, 98C, pp. 14633-14649.

Godfrey, J.S., M. Nunez, E.F. Bradley, P.A. Coppin, and E.J. Lindstrom, 1991: On the net surface heat flux into the western equatorial Pacific. *Journal of Geophysical Research*, 94, pp. 8007-8017.

Goyet, C. and P.G. Brewer, 1993: Biochemical properties of the oceanic carbon cycle. In *Modeling Oceanic Climate Interactions*. Edited by J. Willebrand. NATO Advanced Study Institute, I11, pp. 271-297.

Haines, K., P. Malinotte-Rizzoli, R.E. Young, and W.R. Holland, 1993: A comparison of two methods for the assimilation of altimeter data into a shallow water model. *Dynamics of the Atmosphere and Oceans*, 17, pp. 89-133.

Hartmann, D.L. and M.L. Michelsen, 1993: Large-scale effects on the regulation of tropical sea-surface temperature. *Journal of Climate*, 6, pp. 2049-2062.

Hoge, F.E., A. Vodacek, and N.V. Blough, 1993: Inherent optical properties of the ocean: Retrieval of the absorption coefficient of chromophoric dissolved organic matter from fluorescence measurements. *Limnology and Oceanography*, 38, pp. 1394-1402.

Liu, W.T. and C. Gautier, 1990: Thermal forcing on the tropical Pacific from satellite data. *Journal of Geophysical Research*, 95, pp. 13209-13217 and 13579-13580.

Morrow, R., J. Church, R. Coleman, D. Chelton, and N. White, 1992: Eddy momentum flux and its contribution to the Southern Ocean momentum balance. *Nature*, 357, pp. 482-484.

Peltzer, E.T. and P.G. Brewer, 1993: Some practical aspects of measuring DOC-sampling artifacts and analytical problems with marine samples. *Marine Chemistry*, 41, pp. 243-252.

Roach, A.T., K. Aagaard, and F.D. Carsey, 1993: Coupled ice-ocean variability in the Greenland Sea. *Atmosphere-Ocean*, 31, pp. 319-337.

Shum, C.K., B.D. Tapley, and J.C. Ries, 1993: Satellite altimetry: Its applications and accuracy assessment. *Advances in Space Research*, 13, pp. 11315-11324.

Tapley, B.D., C.K. Shum, J.C. Ries, R. Suter, and B.E. Schutz, 1992: Monitoring of changes in global mean sea level using Geosat altimeter. *Geophysical Monograph*, 69, pp. 167-179.

Walstad, L.J. and A.R. Robinson, 1993: A coupled surface boundary-layer-quasigeostrophic model. *Dynamics of the Atmosphere and Oceans*, 18, pp. 151-207.

Yoder, J.A., C.R. McClain, G.C. Feldman, and W.E. Esaias, 1993: Annual cycles of phytoplankton chlorophyll concentrations in the global ocean: A satellite view. *Global Biogeochemical Cycles*, 7, pp. 181-193.

CHAPTER 4

Greenhouse Gases and Tropospheric Chemistry

Minor species, such as ozone and the hydroxyl radical, have a major influence on the chemistry of the atmosphere and in this capacity have become directly important to human societies.

Earth System Science: A Closer View, 1988

CURRENT UNDERSTANDING

Human activities are changing the concentrations of trace constituents in the troposphere; these in turn affect the radiative balance, dynamics, and chemistry of the atmosphere. One of the most pressing global change issues facing researchers and policymakers today involves the greenhouse effect and potential global warming. Carbon dioxide, nitrous oxide, methane, and CFCs are the principal anthropogenic greenhouse gases that perpetrate this effect. Of these, carbon dioxide has the highest concentration and plays the most significant role. However, the other gases are more efficient at trapping infrared radiation, and their combined effects are almost as large as that of carbon dioxide, even though their concentrations are much smaller. Ozone also proves significant, because the catalytic chemical process involved in its breakdown in the stratosphere and its increasing tropospheric concentrations have a multifaceted influence on the Earth's radiative balance.

The scientific questions become convoluted when considering the sources, chemistry, and ultimate sinks of minor atmospheric constituents and their interrelationships. The planet's inorganic and organic matter are the sources of many of these gases, increasing the importance of land-surface studies to determine emission rates and locations. Yet, the large-scale and seasonal distribution of these sources remains poorly understood. Space-based remote sensing provides the ideal platform for global monitoring of trace gas sources.

Because they control the oxidizing state of the troposphere, concentrations of the hydroxyl and similar radicals strongly influence the chemistry of the lower atmosphere. These radicals are responsible for many of the transformations that occur in the lower atmosphere. Unfortunately, they are extremely difficult to measure, even using ground-based and *in situ* techniques. Therefore, the overall strategy must incorporate the available space-based and *in situ* measurements with databases of gaseous concentrations and emissions, and synthesize them in global climate models to understand the overall problem.

KEY QUESTIONS

- What are the source distributions of radiatively active gases with natural and anthropogenic origins? What is the natural and anthropogenic variability in ecosystems that results from changes in temperature, acid deposition, rainfall, and ultraviolet radiation?
- What is the natural variability in tropospheric ozone? Is tropospheric ozone increasing? Are its precursors (carbon monoxide, nitrogen oxides, and hydrocarbons) increasing? What are the roles of convection, stratosphere-troposphere exchange, and long-range transport in ozone formation? What are the magnitudes of ozone production from industrial activity and biomass burning? What are the effects of ozone changes on forests, agriculture, and ecosystems?
- What are the spatial distributions of the hydroxyl radical and hydrogen peroxide in the troposphere? What is the total abundance of the hydroxyl radical, and is it changing?
- What are the terrestrial and oceanic contributions to the global carbon cycle? What are the anthropogenic contributions to these trace gases, and how are they distributed? Are there trends in the emissions of radiatively and photo-active gases, both natural and anthropogenic? Are there large-scale interactions of clouds and precipitation with trace gases?

SCIENCE STRATEGY

To gain insight into the above issues, we need space-based observations of tropospheric trace gases that have lifetimes of days to months. Researchers have to infer the concentration of short-lived trace gases from ultraviolet flux and measurements of the intermediate lifetime gases, using an appropriate model coupled with data assimilation. Field campaigns can then validate results. Along with trace gas measurements, studies of greenhouse gases also require information from other disciplines (e.g., ocean color, biomass burning, and land vegetation).

Field campaigns and existing satellite data have already revealed relationships between various tropospheric gases and their sources. Currently operating stratospheric instruments such as TOMS and others onboard UARS are helping scientists learn about exchange of trace gases between the stratosphere and troposphere. Differences in ozone concentrations measured from SAGE (stratosphere only) and TOMS (total column) will help approximate residual ozone in the troposphere, providing information on regional and seasonal ozone behavior in the middle troposphere. Prior to the launch of EOS-AM1, surface fluxes of some land- and ocean-derived trace gases will be estimated, using AVHRR and SeaWiFS observations and four-dimensional data assimilation. The Japanese Interferometric Monitor for Greenhouse Gases (IMG) instrument aboard ADEOS also will supply global measurements of long-lived trace gases in the pre-EOS era. In addition, aircraft campaigns add greatly to knowledge of lower stratospheric and tropospheric chemistry. These campaigns focus on transport and transformation mechanisms of specific trace gases, but do not provide global surveys of tropospheric constituents such as ozone, ozone precursors, and other biogeochemically important trace gases.

In the EOS era, TES and MOPITT will investigate the evolution of tropospheric trace and greenhouse gases and their interaction with the climate and biosphere. HIRDLS and MLS will provide additional data on upper tropospheric and stratospheric greenhouse gases, ozone, water vapor, methane, nitrous oxide, and CFCs. These instruments will help address questions of exchange between the stratosphere and troposphere; changes in key greenhouse gases, ozone, and water vapor in the upper troposphere; and changes in the chemistry of the upper troposphere and lower stratosphere, including the role of heterogeneous processes.

Along with the many scientists conducting basic research and developing key observational data products from the instruments mentioned above, several interdisciplinary research teams focus on greenhouse gases and the chemistry of the troposphere and lower stratosphere:

- *Biogeochemical Fluxes at the Ocean/Atmosphere Interface*, Peter Brewer
- *Observational and Modeling Studies of Radiative, Chemical, and Dynamical Interactions in the Earth's Atmosphere*, William Grose
- *Interannual Variability of the Global Carbon, Energy, and Hydrologic Cycles*, James Hansen
- *The Role of Air-Sea Exchanges and Ocean Circulation in Climate Variability*, W. Timothy Liu
- *Changes in Biogeochemical Cycles*, Berrien Moore III
- *A Global Assessment of Active Volcanism, Volcanic Hazards, and Volcanic Inputs to the Atmosphere from EOS*, Peter Mouginis-Mark

- *Using Multi-Sensor Data to Model Factors Limiting Carbon Balance in Global Grasslands*, David Schimel
- *Investigation of the Chemical and Dynamical Changes in the Stratosphere Up to and During the EOS Observing Period*, Mark Schoeberl.

James Hansen of the NASA Goddard Institute for Space Studies oversees a team of researchers investigating the interannual variability of key parameters and processes in the global carbon, thermal energy, and hydrologic cycles. His team uses existing information to establish baseline data sets in anticipation of EOS-era observations. The resultant geophysical data sets allow comparison of output from general circulation models with global diagnostic data, with emphasis on interactions among components of the Earth system, such as land and atmosphere, ocean and atmosphere, and coupled chemistry and dynamics of the atmosphere. These investigators use their general circulation model to assess effects of environmental changes on agricultural ecosystems and food production and on natural ecosystems.

Specifically shedding light into the greenhouse gas issue and possible contributions from tropospheric sources, his team focuses on developing models of global temperature change, which also involves improved quantitative definition of radiative forcing and feedback and of oceanic deep circulation in the transport and exchange of heat, energy, and momentum. Recent findings indicate that cloud feedback can induce decadal variability of mean global temperature, independent of varying ocean dynamics, and a reduction in cloud optical thickness with increasing temperature. Whether this latter relationship holds for the long term is one of the principal global change issues, since it proves central in establishing climate sensitivity.

The *EOS Reference Handbook* (Asrar and Dokken, 1993) contains synopses of the other investigations listed above. For deeper coverage of the science currently pursued by these interdisciplinary science teams, consult the articles cited at the close of this chapter.

EXPECTED RESULTS

Advances in knowledge about the chemistry of the troposphere and lower stratosphere can be expected as a result of interdisciplinary studies that are currently underway and continuing into the next century:

1) **Trace constituent distributions and trends**—High-resolution measurements in the upper troposphere and lower stratosphere will provide global

distributions of many trace gases that affect climate and public health. Observations of aerosols and water vapor will also improve in resolution, coverage, and vertical extent during the EOS era. Analyses of temporal trends in global data sets will extend from the existing and ongoing satellite observations to complete data that include measurements from occultation profiles and total column integrals.

2) **Wind patterns**—Space-based scatterometers will measure winds at the ocean surface, and models—constrained by observations of constituent distributions and physical conditions—will infer wind over land and at higher altitudes. A four-dimensional data assimilation system will provide standard global wind data based on meteorological and constituent inputs. It will estimate transport and mixing rates for atmospheric constituents for use in atmospheric chemistry models.

3) **Exchange between the troposphere and stratosphere**—The effects of trace constituents on temperature profiles and dynamics differ greatly between the troposphere and stratosphere. EOS-sponsored investigations will improve estimates of the vertical exchange of constituents in the tropopause region. They will elucidate mechanisms for transport between the troposphere and stratosphere through process studies that use aircraft and balloons to carry out *in situ* measurements in the atmosphere.

4) **Natural and anthropogenic sources and sinks**—Trace constituents move between the troposphere and Earth's surface, including the biosphere, oceans, volcanoes, and human activities. Biological activity on land and in the oceans drives biogeochemical cycles. Air-sea exchange and chemical transformation of key gases occur in the troposphere. EOS observations and models of biological activity will greatly reduce uncertainty in the components (and their magnitude) of global biogeochemical cycles. EOS investigators will quantify volcanic inputs of gases and aerosols to the atmosphere. Improved access to data sets about human factors, improved models capable of incorporating human influences, and monitoring of those components of the Earth system under the most pressure from anthropogenic activities will address human activities such as the burning of fossil fuels, deforestation, industrial air pollution, agriculture, land-use change, and fertilization of coastal marine ecosystems. Predictive models incorporating a balanced treatment of natural and anthropogenic influences on biogeochemical cycles are a critical policy-relevant need for testing scenarios of greenhouse gas abatement.

5) **Global tropospheric chemistry**—Availability of the information cited above will lead to improved models of tropospheric chemistry, especially for the distribution of global greenhouse gases, ozone, and the hydroxyl radical. Improved knowledge of the distribution and trends for radiatively active gases are critical for improving estimates and predictive models

of the Earth's radiative balance, surface temperatures, and the global hydrologic cycle. Tropospheric chemistry models will also contribute to improvements in modeling regional and local air pollution.

SUGGESTED READING

Bartlett, K.B. and R.C. Harriss, 1993: Review and assessment of methane emissions from wetlands. *Chemosphere*, 26, pp. 261-320.

Fairlie, T.D.A., M. Fisher, and A. O'Neill, 1990: The development of narrow baroclinic zones and other small-scale structure in the stratosphere during simulated major warmings. *Quarterly Journal of the Royal Meteorological Society*, 116, pp. 287-315.

Fung, I., 1993: Models of oceanic and terrestrial sinks of anthropogenic CO_2: A review of the contemporary carbon cycle. In *Biogeochemistry of Global Change: Radiatively Active Trace Gases*. Edited by R. Oremland. pp. 166-189.

Geller, M.A., E.R. Nash, M.-F. Wu, and J.E. Rosenfield, 1992: Residual circulations calculated from satellite data: Their relationships to observed temperature and ozone distributions. *Journal of Atmospheric Science*, 49, pp. 1127-1137.

Liu, W.T., W. Tang, and P.P. Niiler, 1991: Humidity profiles over oceans. *Journal of Climate*, 4, pp. 1023-1034.

Melillo, J.M., D.W. Kicklighter, A.D. McGuire, B. Moore III, C.J. Vorosmarty, and A.L. Grace, 1993: Global climate change and terrestrial net primary production. *Nature*, 363, pp. 234-240.

Ojima, D.S., W.J. Parton, D.S. Schimel, J.M.O. Scurlock, and T.G.F. Kittel, 1993: Modeling the effects of climatic and CO_2 changes on grassland storage of soil C. *Water, Air and Soil Pollution*, 70, pp. 643-657.

Raich, J.W., E.B. Rastetter, J.M. Melillo, D.W. Kicklighter, P.A. Steudler, B.J. Peterson, A.L. Grace, B. Moore III, and C.J. Vorosmarty, 1991: Potential net primary production in South America: Applications of a global model. *Ecological Applications*, 1, pp. 399-429.

Rind, D. and A. Lacis, 1993: The role of the stratosphere in climate change. *Surveys in Geophysics*, 14, pp. 133-165.

Thompson, A.M., W.E. Esaias, and R.L. Iverson, 1990: Two approaches to determine the sea-to-air flux of dimethyl sulfide: Satellite ocean color and a photochemical model with atmospheric measurements. *Journal of Geophysical Research*, 95, pp. 20552-20558.

Tselioudis, G., A. Lacis, D. Rind, and W.B. Rossow, 1993: Potential effects of cloud optical thickness on climate warming. *Nature*, 366, pp. 670-672.

CHAPTER 5

Land Surface:
Ecosystems and Hydrology

Ultimately. . .shifts in the functions of terrestrial ecosystems induce changes in the climate of the globe and the chemistry of the atmosphere and oceans.

From Pattern to Process: The Strategy of the Earth Observing System, 1987

CURRENT UNDERSTANDING

Terrestrial systems have important links to climate and atmospheric composition that are currently absent or inadequately represented in extant coupled models of the Earth system. These links involve exchanges of energy and moisture, radiatively active trace gases like carbon dioxide and methane, and photochemically active species such as methane, nitric oxide, and non-methane hydrocarbons that soils, plants, and biomass burning release. These exchanges depend on properties of the underlying soils and overlying vegetation and on land management practices. Crucial atmospheric inputs include incident radiation, precipitation as water or snow, and atmospheric deposition of nutrients and pollutants, such as sulfur, nitrogen, and ozone. How the surface stores water in its various phases and releases it back to the atmosphere or provides it as a resource to streams and rivers depends on water-holding properties of the soil column and the resistance to water movement exerted by the soil, roots, and especially plant stomata.

Ecologists have proposed, and partly validated, representations of the water and energy exchanges at the land surface in a suitable form for use in global models. We can still improve the process descriptions in these models, but lack of global data, more than understanding of the processes, limits their implementation as part of Earth system models. We lack good global data on surface albedos over land, on the distributions of soils and important properties of vegetation morphology and physiology, and on water storage in soils and snow cover and their relationship to runoff. Incorporation of this information in the land-surface component of climate models will improve estimates of exchanges of water and energy with the atmosphere and lead to more realistic climate simulations.

Representations of terrestrial biogeochemistry in global models are currently more primitive than the schemes used for biophysics. Terrestrial ecosystems are significant sources and sinks for greenhouse gases, and photosynthesis from the land is second only to oceanic uptake in transforming and removing carbon dioxide from the atmosphere. Greater concentration of atmospheric carbon dioxide causes terrestrial vegetation to increase photosynthetic carbon fixation, so vegetation and soils could ameliorate global change by taking up more carbon from the atmosphere. On the other hand, the warmer temperatures that will accompany elevated carbon dioxide concentrations will cause greater rates of respiration and decomposition within ecosystems. Thus, the net response of ecosystems to more carbon dioxide is difficult to assess.

At present, terrestrial ecosystems apparently take up 1 to 2 gigatons per year of the anthropogenic carbon dioxide. Researchers are unable to quantify the role of several mechanisms in creating this sink. Scientists know that about half of the anthropogenic carbon released annually remains in the atmosphere. Of utmost concern is the possibility that this ratio may change in the future. Its estimation requires a better understanding of current sinks, their spatial distribution, and the role of direct human effects in their magnitude. Changes in land use and management, especially deforestation and biomass burning, reduce the standing stock of the terrestrial biosphere and cause the direct release of carbon dioxide into the atmosphere. However, forest regrowth reestablishes the sink, but at unknown rates. The total contribution of deforestation and biomass burning to atmospheric concentrations of carbon dioxide is not known well, and knowledge of the link between these activities and other greenhouse gases is also uncertain.

We also know that atmospheric methane concentrations are increasing at a rate of 0.9 percent annually, but do not know the cause. Several proposed anthropogenic sources for increasing methane concentrations in the atmosphere involve the terrestrial biosphere. These include rice cultivation, biomass burning, and emissions from cattle. In addition, natural wetlands serve as a significant source of methane, and regional change caused by climate warming or increased duration of flooding could lead to large increases in their emissions. Climatic warming and drying of wetlands, on the other hand, could reduce the methane emissions. In areas where the storage of organic carbon in soils is high, warming could lead to increased decomposition and emission of carbon dioxide. Scientists now have a reasonably good understanding of how these processes work in specific regions, but do not understand them on a global scale or over interannual to decadal time periods.

Terrestrial ecosystems will change in response to their biogeochemical environment. Changes in atmospheric concentrations of greenhouse gases are likely to lead to shifts in the species composition, structure, and element cycling in terrestrial ecosystems. These developments may well result in new ecosystems that respond differently to an environment with more carbon dioxide. Therefore, long-term changes and feedback remain unclear.

KEY QUESTIONS

- How will the land surface's biophysical controls on the carbon, energy, and water cycles respond to and feed back on climate change?
- How will climate change affect the respiration component of the carbon cycle, particularly decomposition?
- How will ecosystems respond to climate change and anthropogenic pressure, particularly through feedback into the climate system and cycling and storage of carbon and trace constituents?
- What are the global distributions of atmospheric trace gas sources and sinks, and how are the concentrations changing over time?
- How do land-cover changes and management practices caused by climate change affect the water and energy balances of the land surface?
- How do we best parameterize the heterogeneous biological, biophysical, and biogeochemical processes occurring at large grid scales?

SCIENCE STRATEGY

In this decade, programs such as the International Geosphere-Biosphere Program (IGBP), International Satellite Land Surface Climatology Project (ISLSCP), and GEWEX will examine exchanges of water, energy, and chemicals between the surface and the atmosphere. Investigators involved in the multidisciplinary Boreal Ecosystem-Atmosphere Study (BOREAS) use ground-, airborne-, and satellite-based platforms to collect data on the carbon cycle and biogeochemistry of the boreal forest biome in Canada, its ecological functioning, and associated processes that could contribute to global climate change.

The primary space-based contributors to land-surface studies currently are Systeme de l'Observation de la Terre (SPOT), Landsat, DMSP, and AVHRR. The historical data from the Landsat series are being used to track changes in land cover, land use, and the distribution of biomass. AVHRR observations are the source of global-scale vegetation data, on phenological and sub-seasonal time scales, that are essential for the study of the exchanges of water, energy, and carbon between the land and the atmosphere. Synthetic aperture radars on European, Japanese, and Canadian satellites and the U.S. Shuttle [Shuttle Imaging Radar-C with X-band SAR (SIR-C/X-SAR)] monitor deforestation and surface and hydrological states and processes. The ability of synthetic aperture radars to penetrate cloud cover and dense plant canopies make them particularly valuable in rainforest studies. Their ability to monitor during all seasons makes them indispensable for studies of the high-latitude boreal forest. Monitoring of other surface properties such as soil moisture, regrowth biomass, and plant water will require multifrequency, multipolarization SARs currently not included in the international series.

Starting in 1998, EOS instruments, particularly MODIS and MISR, will support global mapping of surface vegetation so that scientists can model exchange of trace gases, water, and energy between vegetation and the atmosphere. Landsat and ASTER will provide simultaneous multispectral, high-resolution observations to provide finer detail on how the coarser resolution instruments map the heterogeneous surface. In addition, MISR's ability to correct land-surface images for atmospheric scattering and absorption and Sun-sensor geometry will allow better calculation of vegetation properties. MIMR will improve mapping of snow cover and snow water equivalent in all weather, and will help estimate vegetation moisture. MOPITT will provide global measurements of tropospheric methane and carbon monoxide.

Along with the teams of scientists supporting each of the instruments listed above, several interdisciplinary research teams focus on understanding and modeling processes relevant to land-surface hydrology and ecosystems:

- *Long-Term Monitoring of the Amazon Ecosystems through EOS: From Patterns to Processes*, Getulio Batista and Jeffrey Richey
- *Northern Biosphere Observation and Modeling Experiment*, Josef Cihlar
- *NCAR Project to Interface Modeling on Global and Regional Scales with EOS Observations*, Robert Dickinson
- *Hydrology, Hydrochemical Modeling, and Remote Sensing in Seasonally Snow-Covered Alpine Drainage Basins*, Jeff Dozier
- *Use of a Cryospheric System (CRYSYS) to Monitor Global Change in Canada*, Barry Goodison
- *The Hydrologic Cycle and Climatic Processes in Arid and Semi-Arid Lands*, Yann Kerr and Soroosh Sorooshian
- *Global Hydrologic Processes and Climate*, William Lau
- *Changes in Biogeochemical Cycles*, Berrien Moore III

- *Using Multi-Sensor Data to Model Factors Limiting Carbon Balance in Global Grasslands*, David Schimel
- *Biosphere-Atmosphere Interactions*, Piers Sellers.

The interdisciplinary teams have made recent advances in knowledge of land-surface processes. For example, Berrien Moore's team develops global and geographically specific mathematical models and databases that describe ecosystem distribution and condition, the biological processes that determine the flux of carbon dioxide and other trace gases with the atmosphere, and the fluxes of carbon and nutrients to aquatic ecosystems. They have used an ecosystem model to assess the response of net primary production by global land vegetation to a doubling of carbon dioxide, alone and with climate response predicted by atmospheric or coupled general circulation models.

In collaboration with researchers on Moore and Sellers' investigations, David Schimel's team uses models to quantify the effects of temperature changes on global terrestrial carbon storage, through a comparison of three ecosystem models and a simple global database model. Moore's team has also resolved scaling problems for some land ecosystem processes. Interdisciplinary investigations have improved the accuracy of deforestation estimates for Amazonia based on Landsat data, including the accuracy of rates of deforestation, abandonment, and secondary growth. They have also used Landsat data to assess forest damage from industrial pollution in Eastern Europe. Such applications will take advantage of the increased spectral, temporal, and spatial resolution afforded by EOS instruments. Ultimately, the suite of models—within an interactive information system that will integrate a geographic information system, a remote-sensing system, a database management system, a graphics package, and a modern interface shell—will help ecologists assess the impact of natural climate variations and human-induced activities on natural and managed ecosystems.

The *EOS Reference Handbook* (Asrar and Dokken, 1993) contains synopses of the other investigations. For deeper coverage of the science currently pursued by these interdisciplinary science teams, consult the articles cited at the close of this chapter.

EXPECTED RESULTS

The following summaries describe results expected from interdisciplinary studies of land hydrology and ecosystem processes that are currently underway:

1) **Land-surface hydrology**—Newly developed continental- and regional-scale surface hydrologic models include explicit treatment of precipitation,

runoff, soil moisture, and snow and ice dynamics over land. Distributed land-surface models will resolve subgrid-scale hydrological processes, particularly precipitation and soil moisture, within global climate models. Therefore, land hydrology models will contribute to the operation of four-dimensional models to provide continuous global estimates of air temperature, humidity, wind velocity, radiation fluxes, shear stress, cloudiness, snow cover, and precipitation. EOS-supported field studies will provide data sets needed to develop and test scaling effects in hydrological models, and to improve understanding of poorly understood land systems (e.g., arid, alpine, rainforest). Hydrochemical models are also being developed to study the influence of snowmelt and runoff on material transport on land and from land to ocean.

2) **Terrestrial ecosystem dynamics and biogeochemistry**—Multitemporal (sub-seasonal to decadal) measurements of terrestrial vegetation dynamics will allow monitoring of natural, anthropogenic, and climate-induced effects on land ecosystems. Vegetation and soil biota play major roles in global biogeochemical cycles, but the magnitude and natural variability of processes in land ecosystems are poorly quantified. Remote sensing from space currently provides a global view of vegetation, and the EOS Program will organize data sets available from the past decade (specifically AVHRR and Landsat) and will continue and improve observations of the land biosphere. Classification and seasonal monitoring of vegetation types on a global basis allow modeling of primary production and terrestrial carbon balances. Primary production—the growth of vegetation that is the base of the food chain—on land is controlled by soil water availability, soil nutrients, temperature, and radiation flux, along with effects of stresses such as pollutants, drought, and herbivores. EOS investigations are building global data sets for soil attributes, and will provide observations relevant to ion, soil moisture, and light that regulate primary production. Climate data from EOS four-dimensional data assimilation will be crucial for terrestrial analyses. By improving the ability to monitor and model biological activity by plants and soil microorganisms, EOS will contribute significantly to reducing uncertainty in the influence of biogeochemical cycles in controlling climate and atmospheric composition. This aspect of the EOS Program will provide the basis for monitoring the role of ecosystems in controlling atmospheric carbon dioxide concentrations, and validation of models used in making accurate projections.

3) **Land-cover monitoring**—Fine spatial resolution observations of ecosystem extent and character provide a better understanding of land-cover and community-scale changes associated with anthropogenic influences, climate-forced succession, or disturbance. The EOS Program is

assembling existing data sets to analyze decadal land-cover trends detectable from satellite-based observations (e.g., Landsat, SPOT) collected since 1972. Land-cover monitoring on shorter time scales will contribute to effective resource management on regional and local scales. Examples include monitoring crops for efficient irrigation and pest control, forests for health and fire recovery, public lands for good stewardship, urban areas for development patterns, and disaster areas (e.g., floods, volcanic eruptions) for effective relief efforts.

4) **Comprehensive biophysical models**—New models of land-surface physical processes (including hydrology, biology, and meteorology) provide suitable representation of the physical feedback of land processes on Earth's climate. Such models are essential if we hope to improve general circulation models that can assess the effects of future climate change on natural ecosystems, agricultural systems, and most human population centers.

SUGGESTED READING

Dai, A.G. and I. Fung, 1993: Can climate variability contribute to the "missing" CO_2 sink? *Global Biogeochemical Cycles*, 7, pp. 599-609.

Dickinson, R.E., A. Henderson-Sellers, C. Rosenzweig, and P.J. Sellers, 1991: Evapotranspiration models with canopy resistance for use in climate models: A review. *Agricultural and Forest Meteorology*, 54, pp. 373-388.

Holling C.S., 1992: The role of forest insects in structuring the boreal landscape. In *A System Analysis of the Global Boreal Forest*. Edited by H.H. Shugart et al. Cambridge University Press, pp. 170-191.

Kerr, Y.H. and E.G. Njoku, 1993: On the use of passive microwaves at 37 GHz in remote sensing of vegetation. *International Journal of Remote Sensing*, 14, pp. 1931-1943.

Kerr, Y.H., J.P. Lagouarde, and J. Imbernon, 1992: Accurate land-surface temperature retrieval from AVHRR data with use of an improved split window algorithm. *Remote Sensing of Environment*, 41, pp. 197-209.

Moran, M.S., R.D. Jackson, P.N. Slater, and P.M. Teillet, 1992: Evaluation of simplified procedures for retrieval of land-surface reflectance factors from satellite sensor output. *Remote Sensing of Environment*, 41, pp. 169-184.

Parton, W.J., J.M.O. Scurlock, D.S. Ojima, T.G. Gilmanov, R.J. Scholes, D.S. Schimel, T. Kirchner, J.-C. Menaut, T. Seastedt, E. Garcia Moya, A. Kamnalrut, and K.L. Kinyamario, 1993: Observations and modeling of biomass and soil organic matter dynamics for the grasslands biome worldwide. *Global Biogeochemical Cycles*, 7, pp. 785-809.

Penner, J.E., R.E. Dickinson, and C.A. O'Neill, 1992: Effects of aerosols from biomass burning on the global radiation budget. *Science*, 256, pp. 1432-1434.

Potter, C.S., J.T. Randerson, C.B. Field, P.A. Matson, P.M. Vitousek, H.A. Mooney, and S.A. Klooster, 1993: Terrestrial ecosystem production: A process model based on global satellite and surface data. *Global Biogeochemical Cycles*, 7, pp. 811-841.

Richey, J.E. and R.L. Victoria, 1993: C, N, and P export dynamics in the Amazon River. In *Interactions of C, N, P, and S Biogeochemical Cycles and Global Change*. Edited by R. Wollast, F.T. Mackenzie, and L. Chou. Springer-Verlag, Berlin. pp. 123-140.

Schimel, D.S., 1992: Population and community processes in the response of terrestrial ecosystems to global change. In *Biotic Interactions and Global Change*. Edited by P.M. Kareiva et al. Sinauer Associates, pp. 45-54.

Schimel D.S., T.G.F. Kittel, and W.J. Parton, 1991: Terrestrial biogeochemical cycles: Global interactions with the atmosphere and hydrology. *Tellus*, 43, pp. 188-203.

Sellers, P.J., J.A. Berry, G.J. Collatz, C.B. Field, and F.G. Hall, 1992: Canopy reflectance, photosynthesis, and transpiration III: A reanalysis using improved leaf models and a new canopy integration scheme. *Remote Sensing of Environment*, 42, pp. 187-216.

Sellers, P.J., M.D. Heiser, and F.G. Hall, 1992: Relations between surface conductance and spectral vegetation indices at intermediate (100 m^2 to 15 km^2) length scales. *Journal of Geophysical Research*, 97, pp. 19033-19059.

Shimabukuro, Y.E., B.N. Holben, and C.J. Tucker, 1993: Fraction images derived from NOAA AVHRR data for studying the deforestation in the Brazilian Amazon. *International Journal of Remote Sensing*, 15, pp. 517-520.

Skole, D. and C. Tucker, 1993: Tropical deforestation and habitat fragmentation in the Amazon: Satellite data from 1978 to 1988. *Science*, 260, pp. 1849-2024.

Ustin, S.L., M.O. Smith, and J.B. Adams, 1993: Remote sensing of ecological processes: A strategy for developing and testing ecological models using spectral mixture analysis. In *Scaling Physiological Processes: Leaf to Globe*. Edited by J.R. Ehleringer and C.B. Field. Academic Press, pp. 339-357.

Vorosmarty, C., A. Grace, B. Moore III, B. Choudhury, and C.J. Willmott, 1991: A strategy to study regional hydrology and terrestrial ecosystem processes using satellite remote sensing, ground-based data, and computer modeling. *Acta Astronautica*, 25, pp. 785-792.

CHAPTER 6

Ice Sheets, Polar and Alpine Glaciers, and Seasonal Snow

There remains the possibility that, under the influence of global warming, the West Antarctic Ice Sheet might become unstable and surge into the ocean, causing a global rise in sea level . . . .

Earth System Science: A Closer View, 1988

CURRENT UNDERSTANDING

A central concern about global warming is that melting ice sheets will cause the sea level to rise, rendering some coastal areas uninhabitable. A modest (1-m) change in sea level would translate into major shifts in shoreline; even a 30-cm change would be important for coastal communities and coastal engineering practices. The economic and legal ramifications involved with such change would be immense. Determining whether sea-level rise—from a combination of melting ice, increased ocean temperature, and withdrawal of groundwater—is a significant possibility requires input from the oceanography, climatology, hydrology, and solid Earth sciences.

Almost 98 percent of the Earth's water is in the oceans, and most of the remainder is in ice sheets and glaciers. Predictions of sea-level change depend directly on estimates of the mass balance of glaciers on the Earth's land surfaces. Present knowledge of the Earth's second largest reservoir of water is inadequate to assess whether the Antarctic and Greenland ice sheets are growing or shrinking. Changes in the mass balance of glaciers and ice sheets are also time-integrated indicators of climate change. Seasonal snow is the accumulation part of the mass balance equation, and seasonal snow cover is also important for its role in climate and hydrology.

Contemporary observations indicate a modest rate of sea-level change of 1 to 2 mm per year, evidence that the ice sheets are close to equilibrium. However, surface measurements of the ice sheets' mass balances are sparse. Repetitive radar altimeter measurements of the elevation of the ice surface are only available for Greenland to as far north as 72°N (about half of Greenland) and for a small part of eastern Antarctica to 72°S. The accuracy of the radar altimeter over the ice is not good enough to tell us conclusively whether the ice sheets are growing or shrinking, especially at their margins. Results from field work show that some smaller glaciers in the polar and mid-latitudes are growing, but most around the globe are shrinking. A current hypothesis is that this shrinking of the mid-latitude glaciers is the source of water for the observed sea-level rise. However, these extrapolations—from glaciers whose mass balances are monitored—to infer the contributions to sea-level change are prone to error. Where reliable estimates of ice thickening and thinning are available, we do not know why the ice responds to the climate as it does. Moreover, some climate models predict that warmer polar regions would have more precipitation, hence high-latitude glaciers would grow. For all these reasons, we cannot predict the response of the polar glaciers and ice sheets to global warming. Researchers need to determine the present changes in ice mass and to learn more about the rates of snow accumulation, melting, and iceberg calving that accompany global warming.

Measurement of sea-level change also requires knowledge of the vertical and horizontal motions of the Earth's crust along ocean coastlines and near the margins of ice sheets. Space-based observations are needed to establish an absolute vertical reference frame for measurements from tide gauges and altimeters, and to monitor deformations of the solid Earth near the margins of the oceans and ice sheets.

Almost all (98 percent) of the water stored on land as ice is in the vast ice sheets of Antarctica and Greenland, but the remaining 2 percent in alpine glaciers and ice caps is as important, as a source of water for sea-level rise and as an indicator of the spatial variability of climate change. Current estimates show alpine glaciers and ice caps as the major contributor to historical and future sea-level increase. Some models predict a positive mass balance for the Antarctic ice cap due to the increase in precipitation and accumulation with increased temperature in the southern polar region. Seasonal snow cover and alpine glaciers are critical to regional radiation and water balances. Over major portions of the middle and high latitudes, and at high elevations in the tropical latitudes, snow and alpine glaciers are the largest contributors to runoff in rivers and to groundwater recharge. Consistent with global warming in this century, alpine glaciers in most regions have receded, but there are some striking anomalies.

Only a small fraction of the Earth's alpine glaciers and ice caps have been carefully studied.

KEY QUESTIONS

- What are the current mass balances of the polar ice sheets?
- Through what processes are these ice sheets growing or thinning?
- What are the rates of snow accumulation, melting, and iceberg calving, and how do these reveal the key dynamics of ice flow?
- How will the climate system and its changes affect mass balances of alpine glaciers and ice sheets, and what is the resulting effect on sea level?
- Will changes in mass balances of ice sheets and glaciers cause a significant rise in sea level in the next century?
- How will snow cover change as the climate warms; in particular, will a greater fraction of the winter snowfall occur as rain?

SCIENCE STRATEGY

Before the EOS satellite series, airborne synthetic aperture radars, ERS-1 and -2, JERS-1, SIR-C/X-SAR, and Radarsat will provide useful data on extent and dynamics of ice sheets and loss of ice through calving. Radar altimeters designed for ocean topography observations will make some measurements of topographic change over parts of the ice sheets. Measurements of sea level and ice volumes require data on changes in ocean and ice elevation, control of satellite orbits, geodetic measurements of the vertical movements of the land around the oceans and ice sheets, and the establishment of a uniform global reference frame for vertical movements.

The primary EOS instruments capable of contributing to studies of the great ice sheets include GLAS, SSALT, TMR, and DORIS—namely, the payload for the EOS-ALT series, whose first flight is scheduled for 2002. The laser altimeter of GLAS is the central instrument, designed to make mass balance measurements of the polar ice sheets.

MODIS, ASTER, and MISR will provide information on snow cover and glacier features. ASTER, a high-resolution imager on EOS-AM1, will yield data to infer glacier flow. MODIS—on both the EOS-AM and -PM series—will provide broad patterns of spatial extent and coverage. MISR, also on the EOS-AM series, will provide information on snow albedo and its angular variations. High-resolution instruments can then examine areas of change to determine how the climate affects the local mass balance.

In addition to the science teams associated with each of the above-mentioned instruments, several interdisciplinary science investigations examine the polar regions and issues related to ice sheets, glaciers, and sea level:

- *Global Water Cycle: Extension Across the Earth Sciences*, Eric Barron
- *Hydrology, Hydrochemical Modeling, and Remote Sensing in Seasonally Snow-Covered Alpine Drainage Basins*, Jeff Dozier
- *Use of a Cryospheric System (CRYSYS) to Monitor Global Change in Canada*, Barry Goodison
- *Climate, Erosion, and Tectonics in the Andes and Other Mountain Systems*, Bryan Isacks
- *Polar Exchange at the Sea Surface (POLES): The Interaction of Oceans, Ice, and Atmosphere*, Drew Rothrock
- *Earth System Dynamics: The Determination and Interpretation of the Global Angular Momentum Budget Using EOS*, Byron Tapley.

A representative study shows how EOS currently attacks the issue of human-induced global change. Jeff Dozier's team investigates the patterns and processes of water balance and of chemical and nutrient balances of seasonally snow-covered and glaciated alpine watersheds. The Earth's mountain ranges are important components of the global hydrologic cycle, even though they cover a small portion of the Earth's surface area. Snow and ice from mountainous areas account for the major source of the water supply for runoff and groundwater recharge over wide areas of the mid-latitudes; they are sensitive indicators of climatic change; they account for much of the currently observed rise in sea level; and the release of ions from the snow pack is an important component in the biogeochemistry of alpine areas. Acid precipitation is a worldwide problem, and regional and global pollution of the atmosphere with acids occurs largely as a result of human activity. Dozier's team has established that when acid deposition falls as snow, the ionic species concentrate on the outsides of the crystals by snow metamorphism, and an "acid pulse" often occurs in the first flush of spring melt water, when aquatic ecosystems are especially sensitive to acidification. Alpine watersheds are usually weakly buffered, and their acid-neutralizing capacities are not robust; thus, they are vulnerable to increasing acidic deposition.

The *EOS Reference Handbook* (Asrar and Dokken, 1993) contains synopses of the other investigations listed above. For deeper coverage of the science currently pursued by these interdisciplinary science teams, consult the articles cited at the close of this chapter.

EXPECTED RESULTS

During the next decade, several major results are likely from research on glaciers, ice sheets, and seasonal snow cover, including the following:

1) **Prediction of whether global warming leads to an increase in sea level, and the rate of such an increase**—EOS-ALT measurements of ice-elevation change will determine the present contribution of the ice sheets to sea-level rise, thereby helping to constrain the range of estimates for other factors such as thermal expansion of the oceans and groundwater depletion. Monitoring of the ice-surface elevations will provide information on precipitation changes and ice ablation in polar regions, providing essential data for validating estimates of changes in the relationship between climate and mass balance of the ice sheets, which currently only atmospheric models provide. The GLAS laser altimeter will measure mass balance of ice sheets to ±5 percent, which corresponds to a sea-level change of ±0.3 mm/year. For glaciers, ASTER and the synthetic aperture radars will monitor changes in the margins and equilibrium lines of a worldwide set of representative glaciers. In addition, GLAS will monitor changes in the major ice streams and their flow patterns. These measurements will provide critical information on the stability of the West Antarctic Ice Sheet and its possible response to climate warming.

2) **Knowledge of the mass balance of the major ice sheets**—GLAS will measure surface elevations to an accuracy per laser pulse of 10 cm. Repetitive measurements at altimeter orbital-crossover points will detect average elevation changes over 100-km squares to less than 1 cm/year, which typically is less than 5 percent of the average annual mass accumulation. Analysis of the seasonal and interannual fluctuations in these surface elevations will provide information on how variations in precipitation and melting drive decadal scale changes in the mass balance.

3) **Characterization of current and historic snow cover in the Earth's mountain ranges**—The remotely sensed data from MODIS, MISR, and ASTER, along with Radarsat, will measure the spatial distribution of snow properties in a sample set of alpine drainage basins. Field programs in the selected study sites already provide data on snow properties and snow and stream chemistry, and a comprehensive hydrologic model is being developed that includes spatially variable accumulation, melt, and chemical transformation parameterizations to tie precipitation chemistry to stream chemistry. Once models of these processes are verified, they will enable projection of plausible ecological changes that accompany changes in precipitation.

SUGGESTED READING

Bales, R.C., R.E. Davis, and M.W. Williams, 1993: Tracer release in melting snow: Diurnal and seasonal patterns. *Hydrologic Processes*, 7, pp. 389-401.

Blanchard, C.L. and K.A. Tonneson, 1993: Precipitation chemistry measurements from the California acid deposition monitoring program, 1985-1990. *Atmospheric Environment*, 27A, pp. 1755-1763.

Dozier, J., 1992: Opportunities to improve hydrologic data. *Reviews of Geophysics*, 30, pp. 315-331.

Dozier, J. and M.W. Williams, 1991: Hydrology and hydrochemistry of alpine basins. In *Geophysics News 1991*. Edited by D. Knopman and S.A. Morse. American Geophysical Union, pp. 24-25.

Elder, K., J. Dozier, and J. Michaelsen, 1991: Snow accumulation and distribution in an alpine watershed. *Water Resources Research*, 27, pp. 1541-1552.

Foster, J.L. and A.T.C. Chang, 1993: Snow cover. In *Atlas of Satellite Observations Related to Global Change*. Edited by R.J. Gurney, J.L. Foster, and C.L. Parkinson. Cambridge University Press, pp. 361-370.

Goodison, B.E. and A.E. Walker, 1993: Use of snow cover derived from satellite passive microwave data as an indicator of climate change. *Annals of Glaciology*, 17, pp. 137-142.

Marks, D. and J. Dozier, 1992: Climate and energy exchange at the snow surface in the alpine region of the Sierra Nevada, 2, Snow cover energy balance. *Water Resources Research*, 28, pp. 3043-3054.

Nolin, A.W., J. Dozier, and L.A.K. Mertes, 1993: Mapping alpine snow using a spectral mixture modeling technique. *Annals of Glaciology*, 17, pp. 121-124.

Walker, A.E. and B.E. Goodison, 1993: Discrimination of a wet snowcover using passive microwave satellite data. *Annals of Glaciology*, 17, pp. 307-311.

Williams, M.W., A. Brown, and J. Melack, 1993: Geochemical and hydrologic controls on the composition of surface water in a high elevation basin, Sierra Nevada. *Limnology and Oceanography*, 38, pp. 775-797.

Williams, R.S. and D.K. Hall, 1993: Glaciers. In *Atlas of Satellite Observations Related to Global Change*. Edited by R.J. Gurney, J.L. Foster, and C.L. Parkinson. Cambridge University Press, pp. 401-422.

Woo, M.K. and T.C. Winter, 1993: The role of permafrost and seasonal frost in the hydrology of northern wetlands. *Journal of Hydrology*, 141, pp. 5-31.

Zwally, H.J., A.C. Brenner, J.A. Major, and J.G. Marsh, 1989: Growth of the Greenland Ice Sheet: Measurement and Interpretations. *Science*, 246, pp. 1587–1591.

CHAPTER 7

Ozone and Stratospheric Chemistry

The people of the world are conducting a continuing, inadvertent experiment with the Earth's atmosphere that is altering its properties on a global scale.

UARS: A Program to Study Global Ozone Change, 1989

CURRENT UNDERSTANDING

Human-induced chemical changes are occurring in the middle and upper stratosphere at an unprecedented rate, interfering with the natural cycle of ozone formation and destruction. Studies of the Antarctic Ozone Hole have found that ice crystals in stratospheric clouds play a key role in the chemistry that accelerates ozone loss over the South Pole. Similar chemical reactions take place on volcanic particles. Volcanic aerosols make it easier for CFCs to destroy ozone. Researchers determined that local ozone losses at northern mid-latitudes after the eruption of Mt. Pinatubo exceeded 20 percent. Such ozone loss increases the ultraviolet flux into the lower atmosphere, which introduces significant health risks, such as skin cancer and cataracts, and reductions in crop yield.

Current studies validate and refine models of ozone destruction through long-term global monitoring of key stratospheric species and physical processes. Global satellite observations, *in situ* observations from aircraft and balloons, and ground-based methods focus on measurements of the basic physical environment of the stratosphere and a minimum set of species that characterizes the main chemical cycles involving ozone.

The minimum chemical set is determined by a radical and a reservoir from each of the major chemical families (i.e., O_X, NO_X, ClO_X, HO_X) and the source gases for these families (e.g., nitrous oxide for the NO_X family). The measurement set for the physical environment includes temperature, wind, ultraviolet flux, and amounts of clouds and aerosols.

KEY QUESTIONS

- Do we understand the evolution of the compositions of stratospheric trace gases during a period of large anthropogenic changes in the stratosphere?
- Can we clearly separate human-induced from natural change?
- How strongly will chlorine affect the stratosphere during the EOS period?
- How will changes in greenhouse gas concentrations (e.g., methane, carbon dioxide, and nitrous oxide) affect the chemistry and dynamics of the stratosphere?
- How will aircraft emissions of nitrogen oxides in the lower stratosphere affect stratospheric ozone?

SCIENCE STRATEGY

Measurement requirements differ for the various species and physical processes. We need continuous global measurements of temperature, winds, aerosols and clouds, long-lived trace gases, and some radical and most reservoir species, along with long-term, high-precision observations of ozone. Launched in September 1991, UARS currently generates the most important data on stratospheric chemistry. However, UARS has a design life of just 3 years, does not make global measurements of HO_x species, and does not focus on the lower stratosphere—the atmospheric region with the largest uncertainty. The NOAA Solar Backscatter Ultraviolet/2 (SBUV/2) and NASA's TOMS and SAGE II instruments currently monitor the ozone layer. SAGE II also provides high-precision aerosol measurements. The Shuttle-borne Atmospheric Laboratory for Applications and Science (ATLAS) missions provide correlative data for UARS and other long-term programs. Finally, the ground-based measurements made by the instruments that comprise the Network for Detection of Stratospheric Change (NDSC) complement the space-based observations.

Starting with the EOS-AERO1 mission in 2000, in an inclined orbit, SAGE III will continue and improve the high-precision records of ozone, water, and aerosol profiles that SAGE I and II began. Starting in 2003, SAGE III will also fly on EOS-CHEM1 in a polar orbit, as will HIRDLS and MLS, which will observe the upper troposphere and entire stratosphere. HIRDLS will provide, at high spatial resolution, daily global maps of temperature, geopotential heights, 10 key trace species, and aerosols. MLS will complete the required measurement set of trace gases with observations of hydroxyl and halogen oxide radicals. A suite of stratospheric instruments aboard the Environmental Satellite (ENVISAT) series, sponsored by the European Space Agency (ESA), and the continued operation of NDSC will augment the EOS measurements.

Along with the scientists supporting the instruments mentioned above, several interdisciplinary research teams focus specifically on stratospheric chemistry:

- *Observational and Modeling Studies of Radiative, Chemical, and Dynamical Interactions in the Earth's Atmosphere*, William Grose
- *Chemical, Dynamical, and Radiative Interactions through the Middle Atmosphere and Thermosphere*, John Pyle
- *Investigation of the Chemical and Dynamical Changes in the Stratosphere Up to and During the EOS Observing Period*, Mark Schoeberl.

Through a combination of *in situ* and satellite observations and two- and three-dimensional coupled chemistry and dynamic models, these investigations will greatly enhance understanding of exchange of energy and chemical constituents between the troposphere and stratosphere. They will also learn more about the atmospheric processing of trace constituents. A brief summary of Mark Schoeberl's study provides an example. The researchers on this team actively analyze long-term data sets of stratospheric ozone, temperature, and trace gases to improve understanding of atmospheric evolution and to look for inconsistencies and other unexplained phenomena in the data sets. Specifically, this investigation uses UARS data to model the depletion and regeneration of polar ozone. Another objective is to assimilate the trace gases observed by UARS and other satellites into a three-dimensional chemical model to improve existing data sets. This team has focused on developing methods and models of forecasting and data assimilation, with the goal to generate dynamically and chemically balanced global representations of satellite and ground-based observations.

Over the past year, this team has developed a Lagrangian approach to compare observations from satellite (UARS), aircraft (ER-2), balloon (ozonesonde), and ground (Differential Absorption Lidar) with simulations. They have also developed three-dimensional stratospheric transport models and a particle trajectory model to differentiate chemical from dynamic processes. They use these circulation models to assess how environmental changes affect stratospheric ozone depletion. They also develop radiative transfer models to study heating and cooling in the stratosphere, with an emphasis on the effects of polar stratospheric clouds and sulfate aerosols on the radiative balance. They will extend the data record initiated with the Nimbus-7 observations throughout the UARS and EOS missions using forecasts and simulations, providing dynamically and chemically balanced global representations of satellite and ground-based data. These data will significantly improve the evaluation of the balance of trace constituents and meteorological diagnostics, and will help characterize dynamic, chemical, and radiative interactions in the stratosphere.

The *EOS Reference Handbook* (Asrar and Dokken, 1993) contains synopses of the above investigations. For deeper coverage of the science currently pursued by these interdisciplinary science teams, consult the articles cited at the close of this chapter.

EXPECTED RESULTS

Since ozone depletion is one of the more pressing global change issues, researchers are exerting a great deal of effort to determine the distribution, causes, and magnitude of stratospheric ozone concentrations, with specific emphasis placed on the following:

1) **Trend analyses**—Trends in the global distribution of stratospheric ozone elucidate the magnitude, extent, and mechanisms of depletion, beginning with the first TOMS observations in 1978 and continuing through the EOS era. Observations of other constituents and processes relevant to stratospheric ozone chemistry began with the UARS mission and will extend through EOS. Ground observations of ultraviolet radiation flux passing through the ozone layer will be compared with trends in total column ozone monitored from space. Trend analyses play a major role in monitoring the effectiveness of policies initiated in the 1980s to reduce the production of CFCs that harm the stratospheric ozone layer.

2) **Transport and chemical transformations**—EOS investigators model the distribution and mixing of trace constituents in the stratosphere with currently available data. UARS and a series of TOMS instruments monitor seasonal and longer term changes in stratospheric trace gases and aerosols, to provide evidence of chemical and dynamic processes in the stratosphere. EOS observations will drive data assimilation models that can produce a balanced, consistent view of global stratospheric chemistry and dynamics. The atmospheric motion field and transport of the variable trace gases introduce much of the uncertainty in current trend analyses. Assimilation models will infer unmeasured variables, such as stratospheric winds, based on transport of materials.

3) **Aerosols and polar stratospheric clouds**—Observations of the distribution and properties of stratospheric aerosols and polar stratospheric clouds will help evaluate their influence on the ozone layer. Polar stratospheric clouds are sites of enhanced chemical transformations. These ice clouds

form in the lower stratosphere in high-latitude regions under distinct meteorological conditions during polar winters. Process studies will improve understanding of the formation and characteristics of polar stratospheric clouds, and remotely sensed variables are key input for predictive models. Volcanic aerosols are also sites of enhanced chemical transformations. Aerosol concentrations can increase significantly following major volcanic eruptions, which can affect the radiation balance and chemistry of the stratosphere. Monitoring and understanding these natural influences on ozone depletion prove essential in differentiating human-induced perturbations, and help in assessing strategies for adaptation and mitigation.

4) **Solar radiation**—Observations, process studies, and models examine the connection between atmospheric composition, solar inputs, and atmospheric temperature. SOLSTICE II will continue to measure the spectral composition of ultraviolet radiation emitted by the Sun, first captured by SOLSTICE on UARS, to improve understanding of how solar radiation interacts with atmospheric ozone. Total solar irradiance, atmospheric chemical composition, and the distribution of oceans and land combine to determine radiative balance—hence the climate of the biosphere. ACRIM will measure total solar irradiance, the variation of which some researchers believe could offset global warming.

SUGGESTED READING

Bekki, S., R. Toumi, and J.A. Pyle, 1993: Role of sulphur photochemistry in tropical ozone changes after the eruption of Mount Pinatubo. *Nature*, 362, pp. 331-333.

Eckman, R.S., R.E. Turner, W.T. Blackshear, T.D.A. Fairlie, and W.L. Grose, 1993: Some aspects of the interaction between chemical and dynamic processes relating to the Antarctic Ozone Hole. *Advances in Space Research*, 13, pp. 311-319.

Jackman, C.H., A.R. Douglass, S. Chandra, R.S. Stolarski, J.E. Rosenfield, J.A. Kaye, and E.R. Nash, 1991: Impact of interannual variability (1979-1986) of transport and temperature on ozone as computed using a two-dimensional photochemical model. *Journal of Geophysical Research*, 96, pp. 5073-5079.

Lary, D.J., G. Carver, and J.A. Pyle, 1993: A three-dimensional model study of chemistry in the lower stratosphere. *Advances in Space Science*, 13, pp. 331-337.

McIntyre, M.E. and J.A. Pyle, 1993: Model studies of dynamics, chemistry, and transport in the Antarctic and Arctic stratospheres. *Antarctic Special Topic*, pp. 17-33.

Pierce, R.B. and T.D.A. Fairlie, 1993: Chaotic advection in the stratosphere: Implications for the dispersal of chemically perturbed air from the polar vortex. *Journal of Geophysical Research*, 98D, pp. 18589-18595.

Pyle, J.A., G. Carver, J.L. Grenfell, J.A. Kettleborough, and D.J. Lary, 1992: Ozone loss in Antarctica: The implications for global change. *Proceedings of the Royal Society of London B*, 338, pp. 219-226.

Salby, M.L. and P.F. Callaghan, 1993: Fluctuations of total ozone and their relationship to stratospheric air motions. *Journal of Geophysical Research*, 98D, pp. 2715-2727.

Schoeberl, M.R., A.R. Douglass, R.S. Stolarski, P.A. Newman, L.R. Lait, D. Toohey, L. Avallone, J.G. Anderson, W. Brune, D.W. Fahey, and K. Kelly, 1993: The evolution of ClO and NO along air parcel trajectories. *Geophysical Research Letters*, 20, pp. 2511-2514.

Solomon, S., M. Mills, L.E. Heidt, W.H. Pollock, and A.F. Tuck, 1992: On the evaluation of ozone depletion potentials. *Journal of Geophysical Research*, 97D, pp. 825-842.

Stolarski, R.S., R.D. Bojkov, L. Bishop, C. Zerefos, J. Staehelin, and J. Zawodny, 1992: Measured trends in stratospheric ozone. *Science*, 256, pp. 342-349.

Thompson, A.M., 1992: The oxidizing capacity of the Earth's atmosphere: Probable past and future changes. *Science*, 256, pp. 1157-1165.

Toumi, R. and J.A. Pyle, 1992: On the limitation of steady-state expressions as tests of photochemical theory of the stratosphere. *Journal of Atmospheric and Terrestrial Physics*, 54, pp. 819-828.

CHAPTER 8

Volcanoes, Dust Storms, and Climate Change

Since major eruptions . . . are known to influence global climate for years afterward, climate models must include the capability of predicting the occurrence and nature of major volcanic eruptions.

Earth System Science: A Closer View, 1988

CURRENT UNDERSTANDING

Solid Earth processes that influence climate include volcanic eruptions, wind-blown dust, and wind erosion, and all interact with the hydrologic cycle and ecosystems. The injection of particulates and aerosols into the atmosphere can have dramatic geophysical consequences, including increased precipitation, ozone destruction, and the lowering of global temperatures. Ozone destruction associated with volcanic aerosols has received a great deal of attention because of the striking imagery provided by TOMS and MLS instruments currently in space. Moreover, cooling of the middle and upper stratosphere—from the increase in reflectivity caused by the aerosols and the reduced absorption of ultraviolet light because of reduced ozone—plays a significant role in climate change. Whether introduced through biomass burning, volcanic eruption, or industrial activity, aerosols can remain aloft for years and reflect away sunlight beaming toward the Earth, thereby reducing global temperatures. For example, researchers estimate that global temperatures declined by about 0.5°C after the eruptions of El Chichón in 1982 and Mt. Pinatubo in 1991.

Wind-eroded particulates carried into the stratosphere in dust storms will also affect the radiative and chemical properties of the atmosphere and the formation of clouds. Changing patterns of climate and land use either enhance or suppress such sources of dust through increased or decreased aridity and windiness. In contrast to volcanic eruptions, which represent an effect external to the climate system, the interchange of mineral dust between the land surface and the atmosphere represents an interactive process. The high spatial and temporal variability of dust storms makes this interaction a difficult interdisciplinary research issue deserving further study.

The climatic effects of the recent eruption of Mt. Pinatubo in the Philippines are still being evaluated. However, some researchers believe that recent harsh events, including Mississippi River flooding and the blizzards that blanketed the northeastern U.S. the past few years, were a direct consequence of suspended volcanic particulate matter. Mt. Pinatubo shot millions of tons of sulfur dioxide, dust, and ash at least 30 km into the stratosphere. The TOMS instrument onboard Nimbus-7 detected a westward, 7,700-km sulfur dioxide plume just 9 days after eruption. Such violent volcanic events inject millions of tons of ash, gases, and aerosols into the upper troposphere and lower stratosphere over a time scale of a week to several months. Assessing their effect on the Earth system requires estimates of the rate at which solid and gaseous materials are ejected and measurements of the nature and recovery time of the transient disturbances to the atmosphere, land surface, and sea surface. The rate of conversion of sulfur dioxide to sulfate aerosols and the residence times of the aerosols in the stratosphere are responsible for the duration of the regional or hemispheric cooling associated with volcanic eruptions. The direct and indirect effects of eruptions on a wide range of biomes, especially alpine or boreal biomes that are already under stress, can last for decades.

Although explosive eruptions are more spectacular than volcanic activity that produces lava flows and volcanic domes, the less spectacular events release gases into the atmosphere that affect the Earth system. Because sulfur is more soluble in basaltic magmas than in silicic magmas, eruptions of volcanoes such as those in Iceland can inject many times more sulfur into the atmosphere than an eruption of the same volume of silicic magma. The 1783 eruption of Laki in Iceland is a perfect example. Eruption of large volumes of basalt over 7 months resulted in a volcanic fog that affected the weather in northern Europe for at least 2 years.

Volcanoes pose specific requirements to detect and measure the temperatures and morphologies of lavas and plumes. The temporal perspective provided by the EOS sensors is critical to the analysis of both explosive and lava-producing eruptions, and of wind-blown particulate matter.

KEY QUESTIONS

- What are the temperatures and morphologies of lavas and plumes?
- What is the rate at which solid and gaseous materials erupt, and what altitudes do they reach?
- What is the rate of conversion of sulfur dioxide to sulfate aerosols?
- What is the residence time in the atmosphere of volcanic aerosols that affect climate?
- What are the interactions between wind-blown mineral and soil dust and the climate system, and how will these interactions change in the future?

SCIENCE STRATEGY

At present, data from AVHRR, Landsat, SPOT, ERS-1, and SIR-C/X-SAR are used to study the surface features of volcanoes. TOMS, SAGE II, AVHRR, and UARS observations are used to monitor and understand the atmospheric impact of volcanoes. The primary EOS instrument package for surface imaging (i.e., ASTER, MISR, and MODIS) will contribute to studies of volcanic processes, as well as many other solid Earth-related subjects. ASTER, EOSP, and MISR will examine volcanic plumes and aerosols. MLS will provide the vertical distribution of injected sulfur dioxide (even in the presence of volcanic dust). SAGE III will give additional information on aerosols reaching the upper troposphere and lower stratosphere, and will follow their global dispersion and dissipation.

Gas release during an eruption may vary on time scales of a few hours, and the segment of the subsurface magma reservoir that is tapped at different stages of the eruption controls the variation. In lava flow fields, magma production rates, cooling history, and spatial distribution of activity provide important information on the internal structure of the volcano. The high spatial resolution of the EOS sensors is important because of the small size of the phenomena. In addition, the determination of pixel-integrated temperatures requires both high spatial resolution and high spectral resolution. ASTER and Landsat are vital for these temperature determinations. MODIS will detect changes in the areas of the volcanoes, while MISR's ability to determine the amounts of particulates in the atmosphere will be useful in evaluating the effect of volcanoes on climate.

MLS, SAGE III, and TES will determine the concentration of sulfur dioxide and the rate of aerosol dispersion around the globe. Modeling of the dynamics of the eruption plume and rate of release of sulfur dioxide, hydrogen chloride, carbon monoxide, water vapor, and other gases points to the magnitude of the eruption, magma chemistry, and tectonic setting. Monitoring the abundance of gas species—with MLS and TES—accompanying an eruption allows for assessment of magma residence time and tectonic setting.

Along with the scientists associated with the instruments and satellites mentioned above, three interdisciplinary science teams contribute significantly in understanding how volcanoes and wind-blown dust affect global climate:

- *Interannual Variability of the Global Carbon, Energy, and Hydrologic Cycles*, James Hansen
- *Climate, Erosion, and Tectonics in the Andes and Other Mountain Systems*, Bryan Isacks
- *A Global Assessment of Active Volcanism, Volcanic Hazards, and Volcanic Inputs to the Atmosphere from EOS*, Peter Mouginis-Mark.

One EOS interdisciplinary investigation focuses exclusively on the role that volcanoes play in climate change. Directed by Peter Mouginis-Mark, the team examines the physical processes associated with volcanic eruptions, the manner by which volcanic gases are injected into the atmosphere, and the relationship between volcanic eruptions and regional tectonic settings. Currently, team members analyze recent and ongoing eruptions with available satellite, aircraft, and ground-based instrumentation. Existing satellite and *in situ* observations were used to monitor and quantify the materials introduced to the stratosphere by the eruption of Mt. Pinatubo. These were the first estimates ever of volcanic sulfate aerosols from satellite observations. These efforts are crucial for climate modeling, because they provide accurate values of the extent and residence times of volcanic aerosols for input into global climate models. The team has also discovered that eruption plumes are cooler than estimated by their height in the past. Such a revelation lends insight into the mechanisms that drive plume ascent (i.e., not solely heat-driven) and helps with predictions of the dissipation of gases and aerosols over time.

One long-range objective of this investigation is to develop a capability for detecting volcanic activity from space soon after an eruption begins. The benefits include early warning to populations in danger and the ability to mobilize research teams. The team is also developing data sets on the geology of individual volcanoes, including current field observations in Hawaii, Chile, Russia, Guatemala, and Costa Rica.

The *EOS Reference Handbook* (Asrar and Dokken, 1993) contains synopses of the other investigations listed above. For deeper coverage of the science currently pursued by these interdisciplinary science teams, consult the articles cited at the close of this chapter.

EXPECTED RESULTS

Several major results are expected from research on volcanology and wind-blown dust before and during the EOS era, including:

1) **Atmospheric impacts**—Measurements of the extent, temperatures, and height of eruption plume clouds will improve greatly because of the higher temporal and spectral resolution of EOS instruments. Global coverage on time scales of days will allow scientists to study volcanoes worldwide, and to coordinate rapid deployment of aircraft and ground-based instruments for process studies. Better estimates of global emissions of volcanic gases and aerosols significantly contribute to studies of atmospheric chemistry, biogeochemical cycles, and the Earth's radiation budget.

2) **Effects on landforms**—New information on lava flow fields, magma production rates, the cooling history of lava flows, and the spatial distribution of volcanic activity worldwide allows improved estimates of changes in the Earth's surface geology. Monitoring of volcanoes for temperature and surface elevation changes helps predict volcanic activity, and the ability to assess landform change following a volcanic eruption will support disaster relief efforts, management of public lands, and land-use planning.

3) **Climate impacts**—Volcanic eruptions affect global climate through the injection of gases, aerosols, and dust into the atmosphere. The type, amount, and distribution of volcanic materials can be observed remotely from space, and the mechanisms by which these materials are transformed chemically or removed from the atmosphere will be uncovered through coordination of field studies, satellite observations, and model development. The natural effects of volcanoes, which can last for years, will be distinguished from anthropogenic additions of gases, aerosols, and soot to the atmosphere.

4) **Hazard monitoring**—The development of predictive capabilities to determine when and where volcanic activity could endanger human populations will be aided greatly by EOS observations and models. Improved knowledge of landforms and land use in the vicinity of known volcanoes will help plan for evacuation or disaster relief in active regions. EOS investigators will test the potential for volcano alarm systems based on daily observations from space.

SUGGESTED READING

Bluth, G.J.S., S.D. Doiron, A.J. Krueger, L.S. Walter, and C.C. Schnetzler, 1992: Global tracking of the SO_2 clouds from the 1991 Mount Pinatubo eruptions. *Geophysical Research Letters*, 19, pp. 151-154.

Hansen, J., A. Lacis, R. Ruedy, and M. Sato, 1992: Potential climate impact of Mt. Pinatubo eruption. *Geophysical Research Letters*, 19, pp. 215-219.

Hansen, J.E. and A.A. Lacis, 1990: Sun and dust versus greenhouse gases: An assessment of their relative roles in global climate change. *Nature*, 346, pp. 713-719.

Isacks, B.L., 1992: Long-term land-surface processes: Erosion, tectonics, and climate history in mountain belts. In *TERRA-1, Understanding the Terrestrial Environment: The Role of Earth Observations from Space*. Edited by P.M. Mather. Taylor & Francis, London, pp. 21-36.

Minnis, P., E.F. Harrison, L.L. Stowe, G.G. Gibson, F.M. Denn, D.R. Doelling, and W.L. Smith, Jr., 1993: Radiative climate forcing by the Mt. Pinatubo eruption. *Science*, 259, pp. 1369-1508.

Mouginis-Mark, P.J., D.C. Pieri, and P.W. Francis, 1993: Volcanoes. In *Atlas of Satellite Observations Related to Global Change*. Edited by R.J. Gurney, J.L. Foster, and C.L. Parkinson. Cambridge University Press, pp. 341-357.

Mouginis-Mark, P.J. and P.W. Francis, 1992: Satellite observations of active volcanoes: Prospects for the 1990s. *EPISODES*, 15, pp. 46-55.

Mouginis-Mark, P., S. Rowland, P. Francis, T. Friedman, H. Garbeil, J. Gradie, S. Self, L. Wilson, J. Crisp, L. Glaze, K. Jones, A. Kahle, D. Pieri, A. Krueger, L. Walter, C. Wood, W. Rose, J. Adams, and R. Wolff, 1991: Analysis of active volcanoes from the Earth Observing System. *Remote Sensing of Environment*, 36, pp. 1-12.

Realmuto, V.J., K. Hon, A.B. Kahle, E.A. Abbott, and D.C. Pieri, 1992: Multispectral thermal infrared mapping of the 1 October 1988 Kupaianaha flow field, Kilauea volcano, Hawaii. *Bulletin of Volcanology*, 55, pp. 33-44.

Woods, A.W. and S. Self, 1992: Thermal disequilibrium at the top of volcanic clouds and its effect on estimates of the column height. *Nature*, 355, pp. 628-630.

CHAPTER 9

An Overview of International Earth-Observing Capabilities
Planned and Present

The global changes of primary concern to national and international policymakers are those that could impact the Earth's life support system.

Our Changing Planet: The FY92 U.S. Global Change Research Program, 1991

SCIENCE PRIORITIES FOR THE U.S. GLOBAL CHANGE RESEARCH PROGRAM

The purpose of EOS and many of its international counterpart programs is to determine the extent, causes, and regional consequences of global climate change. The extent of global change—for example, the change in average temperature and the time scale over which it will occur—is presently uncertain. The causes can be either natural or human-induced. To explicitly separate the natural events from human perturbations, we must understand and model both natural processes and the influence of anthropogenic contributions. Accurate predictions of regional and global consequences help assess the detrimental effects of climate change and lead to the development of appropriate adaptation and mitigation strategies for those conditions deemed undesirable or unavoidable.

Complex processes within and interactions among the Earth system components—atmosphere, oceans, mantle, biosphere, and lithosphere—and their interactions with the energy input from the Sun determine the behavior and evolution of the climate system. As identified in the previous chapters, there are many important questions about global change. Unfortunately, complexity and other constraints prevent any research program from addressing them all. The EOS

Program considers priorities established by research and policymaking bodies and the science investigators who will analyze the remotely sensed observations as part of the U.S. Global Change Research Program.

The Committee on Earth and Environmental Sciences (CEES) articulated the objectives of USGCRP in a document that accompanied the President's FY91 budget submission (CEES, 1990). Our Changing Planet described the broad priorities that would yield the most relevant knowledge about global change (see Figure 4). The Committee on the Environment and Natural Resources subsumed CEES in January 1994, elevating the visibility of global change research by making CENR one of nine standing research and development coordinating committees under the National Science and Technology Council (NSTC). President Clinton established NSTC in November 1993.

CENR has acknowledged the absolute necessity of the remote sensing from space of certain key Earth system variables, and reaffirmed the science strategy put forth by CEES. This science strategy mandates that NASA support the overall objectives of the broad U.S. and international global change scientific effort, encouraging full participation by the entire Earth science community to the extent that financial limits allow. Emphasis has been placed on global climate change and the interactions between disciplines. Interdisciplinary process studies provide the best opportunity to understand and discriminate between natural and human-induced changes. Knowledge gained from these seven prioritized categories already has yielded integrated conceptual and predictive models, providing policymakers the data and guidance needed to make sound environmental policy decisions (e.g., the Montreal Protocol on Substances that Deplete the Ozone Layer).

The first goals of CEES remain central in defining the EOS Program. However, policy questions and funding restrictions dictate that these broad areas of research be explicitly ranked. The Intergovernmental Panel on Climate Change—an international forum created by the World Meteorological Organization (WMO) and the United Nations Environment Program (UNEP), and endorsed by the U.S. Government—has shaped scientific knowledge of the Earth system into information useful for policymakers. As such, IPCC has targeted several key "scientific uncertainties" that must be addressed to understand how climate changes:

- **Clouds**—Primarily cloud formation, dissipation, and radiative properties, which influence the response of the atmosphere to greenhouse forcing
- **Oceans**—The exchange of energy between the ocean and the atmosphere, between the upper layers of the ocean and the deep ocean, and transport within the ocean, all of which control the rate of global climate change and the patterns of regional change

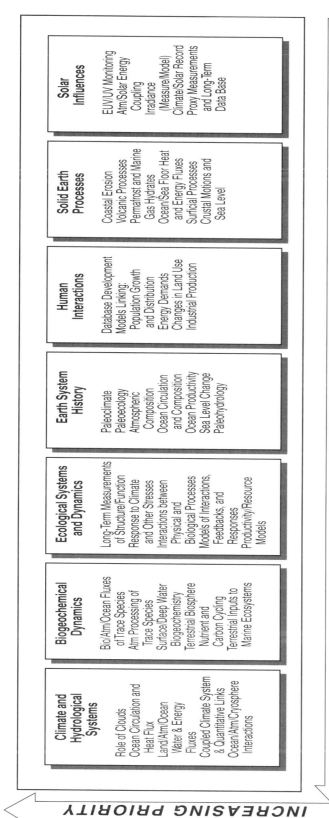

FIGURE 4. U.S. Global Change Research Program Science Priorities

- **Greenhouse Gases**—Quantification of the uptake and release of green-house gases, their chemical reactions in the atmosphere, and how these may be influenced by climate change
- **Polar Ice Sheets**—Predictions of future sea levels and sea state.

IPCC added that land-surface hydrology and the impact on ecosystems are also important realms of study. The uncertainties cited by IPCC—supplemented by studies by other agencies such as the Environmental Protection Agency (EPA), and by the EOS investigators—reinforce the merit and focus of the seven priority research areas identified in Figure 4 as the most important goals for EOS. The following subsections provide programmatic detail to elucidate how the international suite of Earth remote-sensing satellites enhance understanding of these policy-relevant goals.

COMPONENTS OF THE EOS PROGRAM

The EOS Program combines observations and interpretation of data with a scientific research effort. As Figure 5 illustrates, EOS is an information system that provides the geophysical, chemical, and biological information necessary for an intense study of planet Earth. In combination with MTPE Phase I missions, the highly focused Earth Probes satellites, IEOS missions, future geostationary platforms, and supporting ground campaigns of the U.S. and its international partners, the EOS Program will provide a vast library of data and products for access by the user community. EOSDIS will enhance interdisciplinary research and assist in breaking down the intellectual barriers between the traditional disciplines of Earth science by offering an integrated view of environmental data. Successful implementation of the following elements will determine whether the program realizes its potential:

- Creation of an interdisciplinary scientific research program that will support the study of the Earth's climate system, hydrologic cycle, and bio-geochemical cycles
- Acquisition and assembly of a global database of established quality and reliability over a 15-year period—mainly from remote-sensing measurements
- Development of a comprehensive data and information system to serve the needs of scientists from a variety of disciplines studying planet Earth
- Improvement of predictive models of the Earth system, focusing on interactions of system components such as air-sea-land coupling or biological effects on climate.

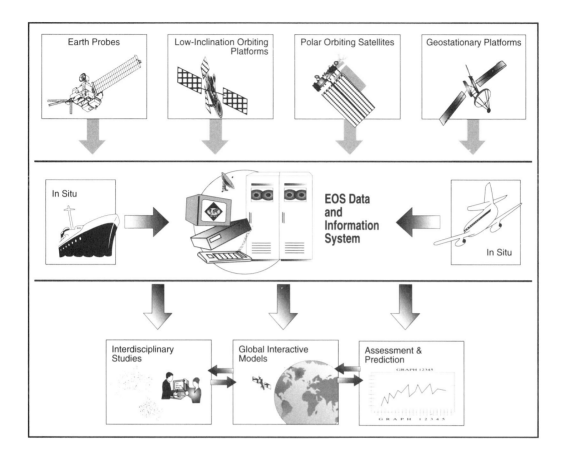

FIGURE 5. EOS and Earth System Modeling

The following subsections delve into the chief components that comprise the EOS Program: 1) Scientific investigations, 2) observational capabilities, 3) a data and information system, and 4) education. Because the previous chapters focus on the science objectives of EOS, the latter three receive emphasis in the coming pages.

The EOS Satellite Series

To provide continuous measurements of global climate change, flights of the principal EOS spacecraft will repeat on about 5- to 6-year centers to ensure adequate coverage for at least 15 years (see Figure 6 for the initial flight scenario). Payloads will change, depending on cost, the evolution of scientific understanding, and the development of technology. To the extent plausible, given platform constraints, synergistic instrument clusters address specific interdisciplinary science problems. In so doing, errors caused by temporal variability in observed phenomena can be minimized. This is particularly significant for removing the effects of the intervening atmosphere from measurements of reflectance or emittance from the Earth's surface, made by passive optical sensors. More information on instrument synergy is available from the *EOS Reference Handbook* (Asrar and Dokken, 1993).

Scientific priorities having been determined and the strategy for addressing them developed, implementation requires launching several suites of instruments into orbit—each designed to accomplish a critical portion of the mission:

- **EOS-AM Series**—Measurement of the diurnal properties of clouds and radiative fluxes and aerosols requires observations in morning and afternoon Sun-synchronous orbits, as well as the inclined orbits provided by TRMM and the EOS-AERO series. In addition, a group of instruments on the morning spacecraft will address issues related to land-atmosphere exchanges of energy, carbon, and water—a task that AVHRR and Landsat address now only qualitatively. Continued acquisition of Landsat data increases the likelihood of success for the EOS-AM series. EOS-AM1 will have an equatorial crossing time of 10:30 a.m., when daily cloud cover is typically at a minimum over land, so surface features can be more easily observed. The instrument complement will obtain information about the physical and radiative properties of clouds (ASTER, CERES, MISR, MODIS); air-land and air-sea exchanges of energy,

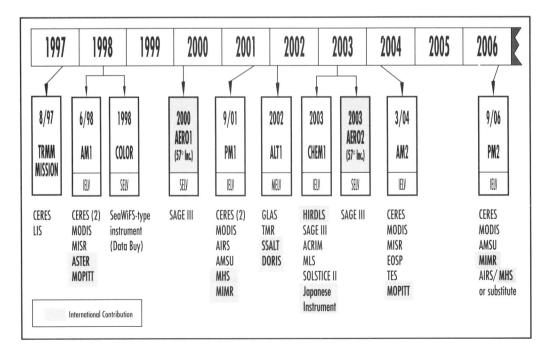

FIGURE 6. EOS Mission Launch Profile

carbon, and water (ASTER, MISR, MODIS); total column measurements of methane (MOPITT); and the role of volcanoes in the climate system (ASTER, MISR, MODIS). The U.S. provides CERES, MISR, and MODIS; Canada provides MOPITT; and Japan provides ASTER.

- **EOS-COLOR Satellite**—The oceans' role in world climate is the second highest priority issue to be investigated by EOS. EOS-COLOR involves a one-time acquisition of observations of the oceans' primary productivity via a data purchase, like that for SeaWiFS scheduled for launch in early 1995. EOS-COLOR will observe ocean color and productivity, with a specific focus on understanding the oceans' role in the global carbon cycle. This mission will provide data continuity until MODIS instruments are flying on both the EOS-AM and -PM series.

- **EOS-AERO Series**—Current plans call for the use of an internationally contributed spacecraft in a 57° inclined, 705-km (or slightly lower) orbit to optimize collection of occultation data in the equatorial and mid-latitude regions. At present, one U.S. instrument makes up the payload—SAGE III, which measures aerosols, ozone, water vapor, and clouds from the middle troposphere through the stratosphere, all important parameters for radiative and atmospheric chemistry models.

- **EOS-PM Series**—This series' afternoon crossing time will enhance collection of meteorological data by the atmospheric sounders onboard. The instrument complement will provide information on cloud formation, precipitation, and radiative properties through AIRS, AMSU, CERES, MHS/AMSU-B, and MODIS. In concert with vector wind stress measurements from a scatterometer (e.g., SeaWinds on ADEOS II), AIRS/AMSU/MHS, MIMR, and MODIS will provide data for global-scale studies of air-sea fluxes of energy, moisture, and momentum. In addition, AIRS, MIMR, and MODIS will contribute to studies of sea-ice extent and heat exchange with the atmosphere. Flight of this platform during the operational lifetime of TRMM will allow assessment of the utility and accuracy of precipitation estimates, with MIMR and MODIS mapping the extent and properties of snow and its role in the climate and hydrological systems. The U.S. provides AIRS, AMSU, CERES, and MODIS; the United Kingdom provides MHS/AMSU-B; and ESA provides MIMR.

- **EOS-ALT Series**—The EOS-ALT payload currently consists of GLAS, TMR, DORIS, and SSALT. GLAS and TMR are U.S. instruments, and France supplies DORIS and SSALT. This payload grouping will ensure continuation of the valuable data set initiated by TOPEX/Poseidon. Investigation of ocean circulation and ice sheets (relevant to sea-level changes) requires accurate altimeter measurements. SSALT is a dual-frequency radar altimeter for the determination of ocean-surface topography, from which ocean circulation can be inferred; SSALT also measures wind speed and wave height. GLAS is a laser altimeter that will generate profiles of ice-sheet volume for Greenland and Antarctica. DORIS enables the precise positioning of the spacecraft (3 to 4 cm), and TMR corrects the altimeter data for pulse delay propagated by water vapor. Plans are underway to separate the radar from the laser to allow flight on separate satellites to support different orbit altitude and inclination requirements.

- **EOS-CHEM Series**—EOS-CHEM instruments will provide measurements of solar energy flux (ACRIM), solar ultraviolet radiation (SOLSTICE II), and atmospheric temperature, aerosols, and gases (HIRDLS, MLS, SAGE III), including the greenhouse gases and the chemical radicals, reservoirs, and source gases that affect ozone depletion. HIRDLS, SAGE III, and MLS—along with MOPITT on EOS-AM1 and -AM2, and SAGE III on EOS-AERO—provide critical data related to tropospheric and lower stratospheric chemistry and dynamics, including troposphere-stratosphere exchanges.

- **Other EOS-Funded Instruments**—Other instruments funded through the EOS Program receive their flight opportunities aboard international partner platforms. CERES and LIS on TRMM will improve diurnal coverage of the tropics, in conjunction with CERES on the EOS-AM and -PM satellites. SeaWinds has been selected as part of the ADEOS II payload; a to-be-determined Japanese instrument (presumably a TOMS-equivalent) will be carried on the EOS-CHEM series.

The EOS science program will combine the measurements from the assembled suite of instruments, using some instruments to correct the measurements of others or to provide data products derived from more than one instrument. This strategy permits the extraction of parameters that a single instrument cannot measure reliably. Global and regional data assimilation models that function as part of EOSDIS will use the observations from the EOS platforms to provide continuous calculations of the principal fluxes between components.

EOS Data and Information System

EOSDIS provides the infrastructure to facilitate interdisciplinary research about the Earth system (see Figure 7). EOSDIS will provide geophysical and biological information, not simply radiance measurements from the instruments. It surpasses the role of a mere data archive; rather, it is a valuable interactive resource of data and derived products readily available to the Earth science community. The data system will operate with an unrestricted data policy, so that anyone can secure

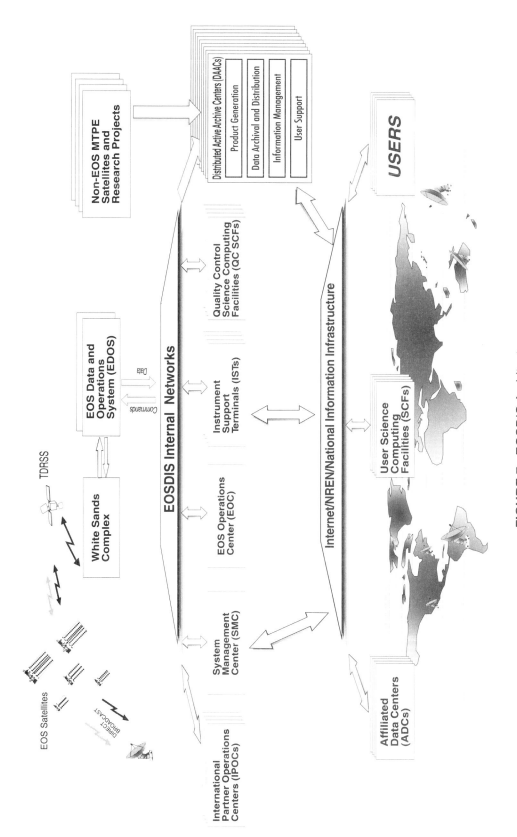

FIGURE 7. EOSDIS Architecture

data for research purposes at a reasonable cost. No preference will be given to EOS investigators, nor will there be any period of proprietary restriction.

NASA is implementing EOSDIS using a distributed, open system architecture. This allows EOS elements to reside at widespread, autonomous facilities to take best advantage of different institutional capabilities and scientific expertise (see Figure 8). Acting in concert, Distributed Active Archive Centers (DAACs) and Scientific Computing Facilities (SCFs) will support global change researchers whose needs cross traditional discipline boundaries, while continuing to support the particular needs of the discipline community. NASA selected eight DAACs covering various types of Earth science data to carry out the responsibilities for processing, archiving, and distributing EOS and related data, and for providing a full range of user support. An additional DAAC—the Consortium for International Earth Science Information Network (CIESIN)—links the EOS Program and socio-economic and educational end users.

EOSDIS is the first element of the EOS Program that will be available to the scientific community, offering useful tools and information at all stages of its evolution. Through a precursor system (Version 0), it will support research and analysis with existing Earth science data, and will establish common protocols for the transfer of data sets. Starting in July 1994, EOSDIS will provide improved access to current satellite data, along with Pathfinder data sets—that is, data from existing operational sensors that have been converted to geophysical and biological parameters in the Version 0 standard, the Hierarchical Data Format (HDF). As EOSDIS evolves, it will build on the capabilities of Version 0. Users' needs for EOSDIS will evolve as researchers work with and respond to the earlier versions of the system. Users' needs will change over time, and EOSDIS will respond in kind.

By the launch of the first EOS platform in 1998, EOSDIS will provide operational capabilities for information management, archival, and distribution. Data from the EOS space component will be processed within a few hours to a few days after observations, depending on the level of data desired [i.e., raw data (Level 0) up to model output (Level 4)]. Researchers will then be able to easily access multiple data sets through an information management and data distribution system, with the networking between DAACs transparent to the end user. The architecture is flexible and resilient, while the implementation will be incremental and evolutionary. EOSDIS includes several distinct but interrelated components:

- **Flight Operations**—Consisting of the EOS Operations Center (EOC) and Instrument Support Terminals (ISTs), this segment provides mission and instrument planning, scheduling, control, and monitoring capabilities.

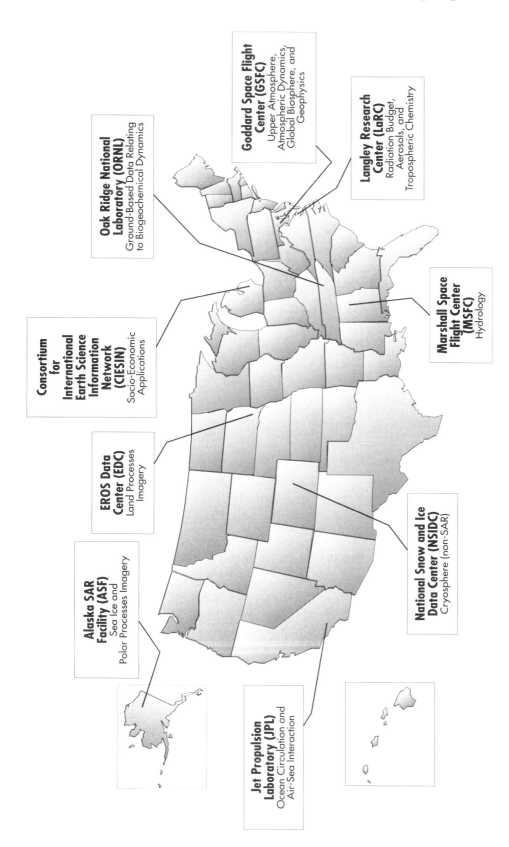

FIGURE 8. EOSDIS-Sponsored Data Centers

- **EOS Data and Operations System**—This system processes Level 0 data, archives them, distributes Level 0 data to the DAACs, and provides a command and telemetry interface to the Tracking and Data Relay Satellite System (TDRSS), which receives the data directly from the spacecraft.
- **Science Data Processing**—This segment consists of the DAACs, which include functions for product generation, archive and distribution, information management, and user support. The primary purpose of this segment is to process higher level data for scientific investigations from measurements taken by the instruments.
- **Communications and System Management**—This segment consists of the System Management Center (SMC) and the EOSDIS internal and external networks. It provides general management and coordination of the ground system resources.
- **Science Computing Facilities**—Located at investigators' sites, SCFs are used to develop and maintain data processing software, to produce special data products, to validate data products, and to perform scientific analyses.

EOSDIS is central to the EOS Program, because it provides the environment in which scientists, educators, policymakers, and other users exploit EOS and other Earth observations data and information. Achieving the goals of the overall mission will depend on how well EOSDIS helps scientists access and use reliable large-scale data sets of geophysical and biological processes under study, and on how successfully EOS scientists interact with other investigators in furthering Earth system science. Currently, progress in use of remote-sensing data for science is hampered by the need for scientists to understand too many operational details. EOSDIS will provide relevant information to a wide range of users, bypassing this operational bottleneck. Scientific data products, created at DAACs and SCFs, will give the user community access to independent measurements to validate and drive models of processes at local, regional, and global scales. EOSDIS will include algorithms for generating the products, descriptors of data quality, and the identity of the responsible scientists.

Four-dimensional data assimilation will play a fundamental role in EOSDIS by integrating the vast number of satellite and conventional observations to provide dynamically and physically consistent data sets for study of the Earth system. Data assimilation enhances the value of the observations by providing a natural mechanism for gridding, quality control, consistency checks, filling in data gaps, and providing estimates of Earth system processes consistent with available observations. Data assimilation represents a natural evolution of simple interpolation and analysis algorithms to further understanding of the Earth system. To be truly useful, observations must be put in context with existing knowledge; the extent that they agree or disagree identifies direction(s) for further research. The four-dimensional data assimilation system being developed for EOS provides a natural vehicle for this process. It will employ a rigorous framework to determine how to best combine model results with asynoptic input to provide contiguous assimilated observations.

Users in the U.S. and participating countries will pay only the marginal costs for data reproduction and distribution, and must agree to publish their results in the open literature and make available supporting information. Users in non-participating countries may have the same access by proposing cooperative projects and associated contributions. There will be no period of exclusive rights to collected data. To the extent possible, the same policies will apply to other relevant data that EOS does not generate, such as observations acquired by other Federal agencies participating in USGCRP or from the international partner platforms.

Education

Over the years, NASA's public outreach activities have assumed myriad forms. The advent of Earth system science now requires an equally innovative program to reach a wider variety of audiences. The mechanism to exploit the educational potential of EOS had to be wholly recreated, because Earth system science goes beyond the existing discipline-specific science curricula offered from elementary through graduate schools. Educators, researchers, and students alike are encouraged and offered incentives to apply the challenging new perspective afforded Earth science.

To develop the science and technical work force needed to collect, interpret, and disseminate the wealth of information provided by EOS and its data system, NASA has undertaken focused efforts to enhance teacher preparation and understanding of interdisciplinary Earth science (MTPE Education Catalog, 1993). Educational materials are developed and widely distributed around the nation to promote the Earth system science curriculum. Enrichment programs start as early as pre-school and continue through grade 12.

Graduate student and postdoctoral research opportunities are also available. Of particular note is the Graduate Student Fellowship in Global Change Research. Established in 1990, these fellowships support graduate students pursuing Ph.D. degrees in Earth system science. The main objective of this program is to train the next generation of scientists and engineers to help analyze and manage the wealth of data and information generated during the EOS era—ingraining an interdisciplinary approach to develop new types of investigators for the maturing global change research community. Fellowships can be renewed up to 2 additional years. NASA has received well over 1,700 applications since the program began, and a total of 284 have been awarded in its first 5 years. The original goal was to scale up to 150 graduate students before the launch of EOS-AM1, but the enthusiastic response from students and faculty alike eclipsed this number. NASA intends to keep the annual scholarship level at about 200 for the life of the EOS mission.

PARTNERS IN THE EOS PROGRAM

Mission to Planet Earth Phase I

Current and near-term NASA Earth science missions directly contribute to global change research before the launch of the first EOS platform—UARS, TOPEX/Poseidon, Earth Probes (i.e., TOMS, NSCAT, and TRMM), SeaWiFS, Landsat, and Space Shuttle payloads:

- Launched in September 1991, UARS carries a payload of 10 instruments that measure the chemical, radiative, and dynamic processes of the stratosphere. The EOS satellites will continue the UARS time series with instruments such as ACRIM, HIRDLS, MLS, and SOLSTICE II.
- The joint NASA and Centre National d'Etudes Spatiales (CNES) TOPEX/Poseidon mission, launched in August 1992, uses radar altimeters to provide precise measurements of broad-scale ocean surface currents, yielding information about ocean circulation and its role in regulating global climate. Copies of the TOPEX/Poseidon altimeter, microwave radiometer, and its orbit determination system are planned for the EOS-ALT series.
- TOMS instruments have proven essential in mapping ozone depletion (particularly the Antarctic ozone hole), and can track the dispersal of sulfur dioxide clouds produced by major volcanic eruptions. International collaborations and a dedicated Earth Probes mission ensure data continuity. TOMS/Meteor-3 is currently operating, and TOMS/ADEOS will launch in February 1996. Plans call for TOMS/EP to launch in late 1994. Discussions are underway for another Russian TOMS collaboration in 1998. SAGE III and TES will continue ozone measurements during the EOS era.
- NSCAT will fly on the ADEOS satellite in 1996, providing information about wind vectors at the oceans' surface. Its follow-on (SeaWinds) is being funded through the EOS Program and is planned for flight on ADEOS II in 1999.

- Scheduled for launch in 1997, TRMM is also a cooperative venture with the Japanese. Two EOS-funded instruments (CERES and LIS) will study clouds, radiation, and lightning in the tropics.
- The SeaWiFS instrument will be launched aboard SeaStar in early 1995, and involves acquisition of ocean primary productivity observations via a data purchase from the commercial operator, Orbital Sciences Corporation.
- The Landsat data archive is one of the most valuable scientific assets available to the Earth science community—the most consistent, reliable documentation of land-cover type and change over the last 2 decades. Scientific integration of Landsat and EOS will greatly strengthen both programs by providing a link to the past and an extension into the future, most notably in the quantification of land-surface parameters that govern energy-water and carbon exchanges. Landsat-7 will launch in late 1998. The EOS satellites will add value to the Landsat data by overlaying high-quality atmospheric forcing fields (e.g., radiation fluxes and humidity profiles) and filling in temporal gaps between the Landsat overpasses with coarser resolution, but more frequent, well-calibrated surface imagery.
- Missions planned for short-term deployment on the Space Shuttle help develop new instruments, prove new technologies, continue vital data sets, and establish calibration for longer term observations. Shuttle payloads that have already flown and will fly again include ATLAS, SIR-C/X-SAR, Measurement of Air Pollution from Satellites (MAPS), and the Shuttle Solar Backscatter Ultraviolet (SSBUV). SSBUV provides cross-calibration of upper atmosphere ozone measurements by comparing measurements with UARS and SBUV/2, while ATLAS observes distributions of trace species throughout the solar cycle. The Lidar In-Space Technology Experiment (LITE) makes its maiden voyage in September 1994.

Many of the instruments that fly before the launch of EOS-AM1 will support large-scale experiments as part of WCRP, IGBP, and JGOFS. For instance, TOPEX/Poseidon now yields essential data for WOCE, and NSCAT will provide data soon. TRMM will provide precipitation data for GEWEX. Coordinated field experiments also provide valuable *in situ* and correlative data. Specially modified and equipped aircraft support field campaigns and help in the development of instruments and algorithms. The MTPE Program folds in contributions from a wide variety of sources. The data collection effort is not solely a space-based endeavor.

The International Earth Observing System

The space- and ground-based elements of the EOS Program are not sufficient to completely sample the dynamic processes of the Earth. Understanding the Earth system and its long-term changes requires a program larger and more expensive than any one nation can manage alone. EOS is part of an international research effort that encompasses space agencies, national and international scientific programs, and researchers, engineers, managers, and policymakers from around the world. The following subsections highlight the contributions of our partners in IEOS—specifically, the Japanese ADEOS I and II, TRMM, and High-Resolution Observation System (HIROS) missions; the European ENVISAT and Meteorological Operational Satellite (METOP) series; and the NOAA POES series.

Japan

Scheduled for launch in February 1996, ADEOS will continue and build on the Earth, atmospheric, and oceanographic remote-sensing measurements that the Marine Observation Satellite (MOS) series and JERS-1 started. The sensors on ADEOS include two core instruments developed by the National Space Development Agency (NASDA) of Japan (AVNIR and OCTS), plus six instruments provided by other agencies and partner nations. CNES will provide POLDER, and NASA will provide NSCAT and TOMS. The conceptual design of an ADEOS follow-on is still underway. One proposal calls for two separate missions— ADEOS II in 1999 and HIROS I in 2000. ADEOS II would carry a global monitoring payload emphasizing the hydrologic cycle, while HIROS I would have high-resolution visible and microwave sensors.

TRMM is a joint NASA/NASDA mission whose primary objective is to measure tropical precipitation—the driver of the hydrologic cycle and atmospheric dynamics. TRMM will measure diurnal variation of latent heating and convective processes in the tropics. Japan provides the Precipitation Radar (PR) and the launch aboard an H-II rocket in August 1997; NASA will provide the spacecraft, the rest of the payload, and instrument integration.

Table 4 lists the instruments slated for the Japanese near-term missions and candidates for the future.

Europe

The ESA contribution to IEOS will be implemented as two spacecraft series— one for environmental monitoring and atmospheric chemistry (ENVISAT), and one for operational meteorological and climate monitoring (METOP). Both satellites will use the *Columbus* polar platform, and both will work with the European Data Relay Satellite System (DRSS).

TABLE 4. Japanese Instruments Planned as Part of IEOS

	Instrument	Category	Measurement Objective
ADEOS	AVNIR	VIS/IR Images	Solar light reflected by the Earth's surface
	OCTS	VIS/IR Images	Ocean color and sea-surface temperature
	ILAS	Stratospheric Chemistry	Stratospheric ozone and related species at high latitudes
	IMG	Tropospheric Chemistry	Carbon dioxide, methane, and other greenhouse gases
	NSCAT	Active Microwave	Wind speed and direction over the oceans
	POLDER	VIS/IR Images and Radiation Budget	Atmospheric aerosols
	RIS	Atmospheric Chemistry	Atmospheric trace gases
	TOMS	Stratospheric Chemistry	Daily global ozone observations
TRMM	CERES	Radiation Budget	Radiation budget
	LIS	VIS/IR Images	Distribution and variability of lightning over the Earth
	PR	Active Microwave	Three-dimensional profiles of rain rates in the tropics
	TMI	Passive Microwave	Precipitation measurements
	VIRS	VIS/IR Images	Variations of rainwater content with altitude
ADEOS II*	AMSR	Passive Atmospheric Sounding and Passive Microwave	Precipitation, water vapor distribution, cloud water, sea-surface temperature, sea ice, and sea surface wind speed
	GLI	VIS/IR Images	Biological/physical processes and stratospheric ozone
	IMG-2	Tropospheric Chemistry	Carbon dioxide, methane, and other greenhouse gases
	SeaWinds	Active Microwave	Ocean surface vector winds
Candidates for HIROS I	ADALT	Altimeter	Geoid, ocean waves, and polar ice
	AVNIR+	VIS/IR Images	Upwelling radiance in multiple spectral bands
	E-LIDAR	Laser Ranging and Sounding	Aerosol and cloud vertical profiles of water vapor, and ice sheet and sea level
	IMB	VIS/IR Images	Ecological environment measurements with high spatial resolution
	PR-2	Active Microwave	Three-dimensional measurement of global rain rate
	SLIES	Stratospheric Chemistry	Infrared emissions of stratospheric and tropospheric minor species
	TERSE	Tropospheric Chemistry	Global monitoring of tropospheric chemistry profiles
	TOMUIS	Stratospheric Chemistry	Three-dimensional mapping of stratospheric ozone distribution
	SAR II	Active Microwave	All-weather monitoring of land/water status

*HIRDLS, ILAS II, POLDER, and TOMS are also candidates for this mission   Core Instruments

ENVISAT has the dual objectives of continuing ERS-1 and -2 measurements (i.e., global ocean circulation and mapping of terrestrial ecosystem type, extent, and phenology) and contributing to environmental studies in land-surface properties, atmospheric chemistry, aerosol distribution, and marine biology. The instrument package includes five ESA-funded core instruments and an additional four from ESA member states. The launch of ENVISAT1 is scheduled for September 1998, with the launch of a second 5 years later.

METOP will fly an operational meteorological package, taking over morning operational satellite coverage from the NOAA system in the year 2000. This collaborative venture with NOAA has the current objective of operational meteorology and climate monitoring, and a future objective of operational climatology. The core operational meteorology payload consists of six instruments, augmented by an additional three. Five instruments have been proposed for climate monitoring. The complete payload will be determined in late 1994, when final program approval and funding are expected. Table 5 lists the instruments and measurement objectives proposed by Europe as contributions to IEOS.

The National Oceanic and Atmospheric Administration

NOAA conducts U.S. civil programs for operational Earth remote sensing. The current and future satellites are an essential part of IEOS, since they provide valuable precursor data and will yield complementary observations during the EOS mission lifetime. NOAA's long-term data record is already being used to establish baseline conditions and to detect trends, with several NASA/NOAA Pathfinder projects underway (i.e., reprocessing of AVHRR, TOVS, and GOES data).

The present POES Program maintains two operational satellites in polar orbit—one providing morning and the other afternoon coverage. NOAA and ESA have agreed that Europe will take over responsibility for the morning global coverage mission in the year 2000, and that NOAA will continue PM coverage indefinitely. Through NOAA, the U.S. will provide a suite of four primary sensors for the METOP mission: AMSU-A1/2, AVHRR-3, HIRS-3, and SEM (see Table 5). Our European partners will reciprocate by supplying sensors and subsystems for flight on both the NOAA morning and afternoon satellites.

Long-term improvements in NOAA satellite, instrument, and space subsystem design will result from technology advances associated with the EOS Program. To this end, coordination in technology development extends to NASA's designating some EOS instruments as "prototypes" for future operational environmental satellites. This means that NOAA and NASA agree on design features that

TABLE 5. European Instruments Planned as Part of IEOS

	Instrument	Category	Measurement Objective
ENVISAT	ASAR	Active Microwave	All-weather imaging of land surfaces, coastal zones, sea ice, and ice- and snow-covered surfaces
	GOMOS	Stratospheric Chemistry	Global ozone monitoring
	MERIS	VIS/IR Images	Ocean color and biological components (carbon cycle)
	MIPAS	Stratospheric Chemistry	Composition, dynamics, and radiation balance of the middle and upper troposphere
	RA-2	Altimeter	Significant wave height and sea-level determination, ocean circulation (dynamics), ice-sheet topography, and land mapping
	AATSR	VIS/IR Images	Sea-surface temperature and land-surface measurements for ocean dynamics and radiation interaction studies
	MWR	Passive Atmospheric Sounding	Corrects the radar altimeter height error caused by water vapor
	SCARAB	Radiation Budget	Global measurements of the radiation budget
	SCIAMACHY	Stratospheric and Tropospheric Chemistry	Total concentration and vertical distribution of atmospheric trace gas species, temperature, and aerosols in the troposphere and stratosphere
METOP	AMSU-A1/2	Passive Atmospheric Sounding	Global temperature profiles from ground level to 45 km
	AVHRR-3	VIS/IR Images	Global monitoring of clouds, sea-surface temperature, vegetation, and ice
	DCS	Data Relay (ARGOS+)	Relays data from over 4,000 collection platforms worldwide
	HIRS-3	Passive Atmospheric Sounding	Global atmospheric temperature profiles, atmospheric water content, and cloud properties
	IASI	Passive Atmospheric Sounding	Atmospheric temperature and humidity profiles, and continuous monitoring of global radiation, dynamics, and energy flux
	MHS	Passive Atmospheric Sounding	Precipitation and water vapor profiles
	ASCAT*	Active Microwave	Ocean surface wind speed and direction, and ocean circulation and dynamics
	AATSR*	VIS/IR Images	Sea-surface temperature and land-surface measurements for ocean dynamics and radiation interaction studies
	GOMI*	Stratospheric Chemistry	Global ozone distributions from both solar radiation backscatter and differential optical absorption spectroscopy
	MCP	Data Relay	Direct data handling and broadcast of operational instrument data streams to ground stations
	MIMR*	Passive Microwave	Precipitation rate, cloud water, water vapor, sea-surface roughness, sea-surface temperature, ice, snow, and soil moisture
	S&R	Data Relay	Receives beacon signals, and transmits in real-time to ground stations around the world
	SCARAB*	Radiation Budget	Global measurements of the radiation budget
	SEM	Other	Monitors particles and fields to measure/predict solar events

*Proposed climatological monitoring instruments

Core Instruments

would enable these prototypes to be transferred to NOAA spacecraft after being space-proven by NASA research and demonstration.

SUMMARY

EOS is a comprehensive, multifaceted program designed to realize the full promise of Earth system science and the global change research effort. The EOS Program combines space- and surface-based data collection and analysis with computer models of the Earth's interrelated processes, completing a logical progression of Earth science endeavors begun over 20 years ago. EOS and partner Earth science missions now assume an even greater significance in that they provide a new level of information and awareness about the Earth. By the end of this century, a powerful international observing system will be in orbit. This system will provide remarkably broad coverage of the Earth in a wide variety of spectral bands, both optical and microwave, with active and passive techniques used in an unprecedented effort to understand how the Earth functions as an integrated system.

CHAPTER 10

Assessing the Impact
of Climate Change

It is only in the moment of time represented by the present century that one species has acquired the power to alter the nature of the world.

Silent Spring, 1962

ACCELERATION OF CLIMATE CHANGE

The Earth's climate changes over time. In years past, natural scientists depended on physical evidence such as tree rings, the stratification of rocks, and ice cores to trace climate change over extended periods. These paleoclimate indicators show many fluctuations, including some instances of substantial changes in climate over a few decades. However, the paleoclimate indicators lack some essential clues about the causes of climate change. Today, we track the processes that contribute to global change with space-based, remote-sensing technology.

Human activity has influenced the environment since the arrival of the species. Through the 18th century, however, the effect was local, limited to regions next to pockets of human settlement. The Industrial Revolution and concurrent population explosion of the 19th century extended the human dominion to the entire planet. Local effects vaulted to global impacts, stressing the Earth system in ways that we have yet to fully comprehend. Humanity's rise has altered the Earth's natural evolution. We have become an agent of change, but we need to evaluate how and to what extent. Some studies indicate that by the middle of the 21st century humans may cause a shift in climatic patterns that would eclipse the irregular variations of nature over recorded history. Increases in resource use, energy consumption, food requirements, the need for freshwater, the spread of urban communities, and the crowding out of natural flora and fauna all stem from a burgeoning human population.

The habitability of the Earth is no longer assured. Our survival and that of other species and natural and managed ecosystems depend on our life support system remaining intact. The Earth provides certain boundary conditions within which millions of species flourish; tinkering with its disparate elements could cause us to unwittingly limit the robust world in which we live. Current models indicate a likelihood of this scenario, but many of the input data are poorly characterized and subsequent model results suspect. Researchers lack the long-term, consistent measurements of key biological, chemical, and physical variables that help define changes in state and modes of variability within the Earth system. Only with these data can we extract human effects from natural processes. If we cannot accomplish this fundamental objective, the climate change debate could continue unabated, with formulation of environmental policy at risk.

As the preceding chapters have indicated, scientists are beginning to understand how the Earth functions as a system. Remote-sensing technology contributes significantly to this enhanced knowledge and gets further refined with each succeeding generation of sensor. Through informed decisions by world leaders based on data provided by the scientific community—system developers, interdisciplinary investigators, global change modelers, and science policy advisors—we will be able to preserve our home planet and sustain our quality of life (see Figure 9).

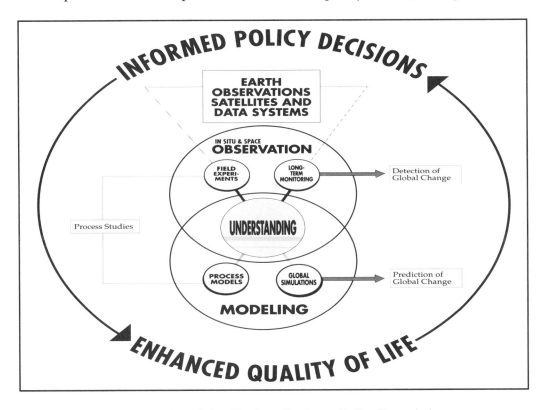

FIGURE 9. The Scientific Contribution to Policy Formulation

AWARENESS OF CLIMATE CHANGE

The increased awareness propagated by scientific discovery and the mass media makes issues such as ozone depletion, global warming, decreasing biodiversity, poor water quality, and land-use degradation a common concern, no longer of interest only to scientists. Recent droughts in the western and southeastern U.S., powerful hurricanes in Florida and Europe, and substantial flooding in the midwestern U.S. and Europe represent the extreme events that may occur with greater frequency if climate changes. Predicted repercussions include increased temperatures, rising sea levels, changing patterns of precipitation, and increased evaporation. Academia, the private sector, and governments around the world recognize the importance of confronting such global problems; however, the general public ultimately drives response strategies.

Most people now realize that they are part of a world population, which must respond collectively to global change issues. Global change researchers hope to understand the processes that we initiate unintentionally, and inform the public of plausible consequences. Currently, most people understand that increasing concentrations of anthropogenic trace gases will cause a rise in global temperatures, but the more troubling, less publicized issue of redistributed precipitation is even more important. Altered precipitation patterns could lead to more floods or droughts, and the disruption of water supplies. Severe droughts could lead to forest fires and the loss of wetlands in the interior of continents. This in turn could result in migration of animals and plants toward cooler and moister climes. Shifts in agricultural and forest ecosystems would influence the demography of rural and urban communities. This simple case study indicates that apparently disjoint problems, such as rainfall and urban development, are indeed related, making the job of global change modelers extraordinarily demanding. The magnitude of the policymakers' task is huge.

The U.S. Government currently works with governments worldwide to address global change issues. Since the 1992 United Nations Conference on Environment and Development (UNCED), more than 190 countries have signed the Climate Convention Protocol, seeking to freeze annual greenhouse gas emissions at 1990 levels. On April 23, 1993, the U.S. announced plans to accelerate the timetable and return to 1990 levels by the year 2000. The U.S. has signed international agreements such as the Montreal Protocol to reduce emissions of stratospheric ozone-depleting chemicals, the international Framework Convention on Climate Change, and the Convention on Biodiversity. In addition, efforts are underway to develop international consensus on desertification and the protection of forests. The release of the U.S. National Climate Change Action Plan (October 1993), on how the U.S. intends to accomplish the reduction of greenhouse gas emissions, shows that decisionmakers are ready to act on the recommendations of the research community, and that they take seriously their responsibility to safeguard the environment.

Of course, the policy responses to identified climate change issues need to be weighed against such concerns as economic vitality, regional stability, and many other factors. Two possible paths come to mind: 1) Rely on our ability to adapt to climate changes regardless of their heritage, and 2) mitigate—through environmental policy decisions, regulation, and technology innovation—the known processes that are causing climate change. The first option requires better predictive models. The second requires scientific understanding of the spectrum of factors that contribute to change before identifying a course of action. To adopt either or combinations of the two, one has to have reliable, timely access to the necessary information. The complexity and scope of environmental issues dictate comprehensive treatment through an interdisciplinary and international cooperative approach.

THE GLOBAL PERSPECTIVE AFFORDED BY EOS

Decisionmakers need an accurate picture of the state of the Earth and the role that humanity plays in altering its physical makeup. Climate change can adversely affect the current human condition and the plant and animal life with whom we share the planet. To support the national and international decisionmaking process, researchers need to differentiate human-induced change from natural change, gauge the severity of the anthropogenic perturbations, and determine what can be done to offset unintended harm. This process is already underway.

EOS-funded interdisciplinary investigators use existing satellite data sets, *in situ* measurements, and observations provided by Earth remote-sensing satellites now in orbit to further their research, with findings disseminated immediately to the public. Recent work at Goddard Space Flight Center, the University of Maryland, and the University of New Hampshire serves as an excellent example of science influencing policy. EOS scientists have used Landsat, SPOT, and AVHRR imagery to monitor deforestation in Amazonia. Concerns over tropical deforestation arise because of several global implications, most notably the quantities of carbon dioxide and other radiatively active gases released to the atmosphere from burning and the potential destruction of ecosystems by fragmentation. The short-term, local benefits associated with such land use (i.e., tree harvests, farming, and grazing) had to be extrapolated to reveal long-term, global implications in a manner acceptable to both the international community and the Brazilian government. A study based on Landsat Pathfinder work (Skole and Tucker, 1993) provided a more reliable estimate of deforestation than previous methods based on local inventories. What is more important, their research showed a significantly lower rate than most other estimates. All parties endorsed their satellite technique to gauge future change.

International agreements directly result from researchers making their findings widely available. For instance, UARS and TOMS observations and airborne and *in situ* measurements have vividly documented an alarming trend in ozone depletion, and global decisionmakers acted on the science community's advice by implementing the Montreal Protocol. Global change remains visible, and awareness of ozone depletion, global warming, deforestation, and other changes will increase soon. These identified risks will continue to receive attention, but what about the as-yet-undiscovered threats to human health and safety? The EOS Program will help provide the global perspective needed to monitor the Earth system variables deemed most significant by the science community. At present, NASA focuses on five policy issues—identified by the national and international scientific community—to hone EOS measurement objectives:

1) Ozone depletion in the stratosphere, resulting in a significant increase in the ultraviolet radiation reaching the Earth's surface, which could cause considerable health hazards
2) Climate variability caused by natural and human-induced activities that could affect patterns of precipitation and temperature, thus agricultural and industrial production and distribution
3) Global warming and long-term climate change, which could contribute to diminished water supplies of suitable quality for agricultural and industrial use, and can be traced to deforestation, other anthropogenic changes to the Earth's surface, and human-induced introduction of trace gases to the atmosphere
4) Decline in the health and diversity of animals and vegetation because of long-term changes in atmospheric chemistry, precipitation, runoff, and groundwater
5) Social and economic consequences of climate changes and their effects on health, standard of living, and quality of life.

The EOS Program responds to contemporary concerns and potential unexpected problem areas that may emerge over the life of the mission. Tables in the following subsections identify specific EOS instrument contributions to the policy issues mentioned above. Though not exhaustive, they provide the reader a sense of the capabilities that EOS provides, the multiple applications of remotely sensed data, and the overarching research questions that help define the program.

Atmospheric Ozone and Ultraviolet Radiation Hazards

The development of the Antarctic Ozone Hole and the global depletion of stratospheric ozone have been and continue to be well-documented by satellite and *in situ* observations. Serving as the primary shield of harmful ultraviolet radiation and a significant element of the global energy balance, the depletion of stratospheric ozone can be tied to two major causes: 1) A six-fold increase in stratospheric chlorine above natural levels, in large part due to the emission of industrial chemicals such as CFCs; and 2) the volcanic emission of sulfur dioxide, which can lead to interannual variability. For example, the eruption of Mt. Pinatubo caused a 35 percent reduction in stratospheric ozone at low northern-hemisphere latitudes. Many other factors influence ozone chemistry. Ozone depletion can be caused by other radical species that are products of industrial emissions, and the emissions from high-altitude aircraft can also perturb stratospheric chemistry.

The EOS Program will contribute substantially to the detection and prediction of ozone depletion, the isolation of human versus natural factors initiating changes in atmospheric ozone, the response of atmospheric ozone to human forcing, and understanding of the uncertainties associated with ozone depletion and ultraviolet-B (UV-B) radiation (see Table 6). Long-term, well-calibrated measurements of ozone and other critical gases, temperature, and aerosols by EOS and other MTPE missions prove critical in differentiating natural variations from human-induced ozone depletion. The EOS global vertical sounding capabilities for the primary radicals, reservoirs, and source gases in the stratospheric chemical cycle and EOS aerosol measurements enable predictive and diagnostic models for ozone depletion and UV-B radiation flux. EOS investigators have developed three-dimensional chemical transport models of the stratosphere, with predictive and diagnostic modes, which can calculate trace gas concentrations, estimate ozone losses, and calculate UV-B. The knowledge from these efforts will continue to guide policy decisions about the use of CFCs and possible industrial refrigerant substitutes.

Climate Variability and Economic Development

Short-term, effective climate forecasting definitely would yield social and economic benefits. For instance, the skillful prediction of statistics about precipitation and temperature months to years in advance would augment agricultural and forestry production and distribution, improve hydroelectric power planning and utilization, and help temper flood and drought effects. Some aspects of the climate system can already be predicted in the short term (e.g., El Niño events). With a comprehensive knowledge of atmospheric circulation and its interaction with the ocean and the land surface, longer term forecasting and extrapolations of

TABLE 6. Atmospheric Ozone and Ultraviolet Radiation Hazards

global change ISSUE	priority research QUESTIONS	relevant EOS INSTRUMENTS	key CONTRIBUTIONS
Atmospheric Ozone and Ultraviolet Radiation Hazards	What are the human and natural forcing agents affecting stratospheric ozone concentrations?	AIRS MISR MLS SAGE III TES	Accurate record of atmospheric and surface temperature fluctuations Total column abundance of aerosols Anthropogenic chlorine and other known threats from industrial chemicals, and volcanic gases (not degraded by clouds) Anthropogenic chlorine and other known threats from industrial chemicals, and polar stratospheric clouds and temperatures Volcanic gases
	What has been and will be the response of atmospheric ozone and UV-B radiation to human forcing?	MLS MODIS SAGE III	Chemical radicals, chemical reservoirs, source gases, and volcanic injections Total column ozone twice daily Stratospheric ozone measurements, as well as NO2, nitrate radicals, ClO2, water vapor, polar stratospheric clouds, temperature, and aerosols
	What are the uncertainties and the potential for change?	CERES MLS MISR SAGE II	Cloud forcing products useful for diagnostic measurements of UV-B transmission Early-warning capability via vertical profiles of chlorine, additional trace gases, and constituents involved in as-yet-undiscovered stratospheric processes Cloud forcing products useful for diagnostic measurements of UV-B transmission Stratospheric temperature fluctuations (potentially induced by increasing atmospheric CO2)

the effects of greenhouse gases are possible. Many of the challenges inherent to long-term prediction are possible on seasonal to interannual time scales. However, an enhanced understanding of natural variability is necessary to give long-range forecasts of global change models a higher degree of confidence. Such knowledge would help distinguish human-induced climate change.

To provide a sound basis for national policy development, two capabilities are being pursued as part of the EOS Program: 1) Operational predictions of interannual climate fluctuations up to 1 year in advance, and 2) the detection, causes, and impacts of natural climate variations on decadal scales. Table 7 lists some of the key EOS instruments involved in realizing these goals.

EOS observations will help define the natural and human influences affecting seasonal and interannual climate variability. Global sea-surface temperature, varying land moisture and vegetation, and the global effects of land, snow, and sea ice cover all contribute to atmospheric circulation, which significantly affects natural variability. Solar radiation and volcanic eruptions also play major roles. Human-induced modifications to atmospheric chemistry may alter the nature of seasonal to interannual variability. Studies are already underway to meld regional satellite and field measurements of hydrologic parameters sensitive to climate change with historical observations. These data assimilation efforts provide

TABLE 7. Climate Variability

global change ISSUE	priority research QUESTIONS	relevant EOS INSTRUMENTS	key CONTRIBUTIONS
Climate Variability and Economic Development	What are the natural and human factors that affect the seasonal to interannual climate and its anomalies, and what is the mechanism for verifying consequent predictions?	ACRIM AIRS AMSU CERES EOS-COLOR EOSP GLAS MHS MIMR MISR SAGE III SSALT	Total solar irradiance to help determine the Earth's radiative balance Accurate record of atmospheric temperature fluctuations Humidity measurements to help assess the role of the land surface in governing seasonal to interannual climate variability at global to regional scales Earth's radiation budget and cloud properties, thus cloud-climate interactions Role of oceans in the global carbon cycle, fluxes of trace gases at the air-sea interface, and ocean primary productivity Total column aerosol optical thickness, and particle sizes and refractive index Growth/retreat of glaciers and ice sheets Atmospheric temperature measurements Precipitation rate and soil moisture Planetary and surface albedo Sulfate aerosols from volcanic eruptions, which can influence climate for years Sea-surface and polar ice-sheet topography to help discern air-sea interactions

baselines for determining patterns of climate variability. EOS-sponsored interdisciplinary investigations are building global models to address this issue—ultimately to project climate variability seasons to decades in advance. This predictive knowledge could help mitigate the social and economic effects of such variability on the agricultural and industrial sectors.

Global Warming and Long-Term Climate Change

Human activities have altered the land surface and the composition of the atmosphere so that long-term climate change may be inevitable. Without a better understanding of the magnitude and rate of such change, potential societal impacts cannot be evaluated. Deforestation and other anthropogenic changes to the Earth's surface can affect the carbon budget; patterns of evaporation, precipitation, and soil erosion; and other components of the Earth system. Land-surface changes modify energy-water-vegetation interactions at the land-atmosphere interface, and biophysical models of this process indicate significant effects on regional climate. Increases in carbon dioxide and other radiatively active gases influence the global energy budget and ultimately promote the greenhouse effect, yet global warming still remains the subject of considerable debate. The observing system in place over the last 100 years has proven inadequate to isolate long-term natural variability from anthropogenic change. Researchers can detect subtle changes, but do not have the precision to determine exact origins. The EOS Program will provide the needed calibration and precise measurements to determine sources of minute global temperature fluctuations.

The detection of climate change, reduction in the uncertainty associated with climate change, the improvement of estimates of global warming potential and sources and sinks of greenhouse gases, and improved prediction of regional change (including natural variability) depend on an adequate monitoring capability, improved knowledge of the forcing factors, an improved predictive capability to determine the response of the climate system, and an assessment of the major uncertainties. EOS will provide unprecedented opportunities to document surface fluxes of energy and moisture over the land surface and the oceans, to track the regional and global radiation budgets required to understand cloud-climate feedback, and to determine ocean heat transport. In addition, EOS will provide the first opportunity to examine ice-sheet mass balance. Lack of measurements and understanding of the underlying processes affecting these variables limit the accuracy of current general circulation models.

EOS observations of the temporal and spatial variability of aerosols, vegetation, temperature, water vapor, ozone, and clouds will better define natural variability of the climate system (see Table 8). With these data, EOS investigators will be able to assess variability resulting from cloud interactions, vegetation changes, perturbations of carbon and trace gases, and the major climate forcing factors. EOS investigators are performing extensive model experimentation aimed at quantitative comparison of the different human-induced and natural forcing changes, with sensitivity experiments targeting both regional and global scales. EOS four-dimensional data assimilation models will provide a physically consistent global measure of climate observations, and planned improvements to climate models will also contribute to more realistic simulations of future climate scenarios, which can be verified by the long-term EOS observations. By exploiting these data, decisionmakers will be better able to assess the economic tradeoffs of adaptation versus mitigation in establishing environmental policy.

Ecosystems and Biodiversity

Human activity is responsible for the mass extinction of countless species, a practice that we must reverse if we hope to fulfill our role as stewards of the global environment for future generations. Reduced biodiversity influences ecosystem structure, including predation and dominance, and diminishes the natural resources available to us. The historical record indicates dramatic expansion and contraction of habitats in response to past global climate changes. Today, human activities play a considerable role in aiding or hindering habitat response. Human activity has greatly accelerated change to the point that many species cannot adequately cope. For instance, sea level increases in response to global warming could dislocate estuaries, marsh, and near-shore

TABLE 8. Greenhouse Warming and Long-Term Climate Change

global change ISSUE	priority research QUESTIONS	relevant EOS INSTRUMENTS	key CONTRIBUTIONS
Greenhouse Warming and Long-Term Climate Change	What are the human-induced and natural forcing changes in the Earth system?	ACRIM ASTER MISR MLS MODIS MOPITT SAGE III TES	Solar variability studies, in tandem with EOS-collected cloud formation/dissipation data Changes in surface cover and vegetation revealing human land-use patterns High spatial resolution images of land cover to document human impact Assessment of the role of volcanism as natural forcing factor Abundances and trends of ozone depletion due to industrial activities CO_2 and CH_4 measurements to isolate human from natural greenhouse forcings Natural vs. human-induced perturbations to global atmospheric chemistry Tropospheric pollution and troposphere-stratosphere exchange
	What has been and will be the response of the climate system to human forcing?	ACRIM AIRS ASTER CERES EOSP MHS MIMR MISR MLS MODIS SAGE III	Total solar irradiance variation as a forcing factor Records of atmospheric temperature fluctuations Land-surface change and radiative forcings (e.g., cloud amount, type) Role of clouds and Earth radiative balance Anthropogenic aerosol loading and impact on regional and global climate Atmospheric temperatures to document rate and magnitude of global warming Ice and snow cover and their role in climate system Impact of anthropogenic aerosols (e.g., industrial emissions, slash-and-burn agriculture, desertification) on global radiation budget Improved basis for global warming potential for identified greenhouse gases Regional sources/sinks for CO_2 in terrestrial and oceanic ecosystems Aerosol global warming potential, and their interactions with anthropogenic chemicals
	How do human-induced changes compare to variations in the natural system, and can these changes be detected?	ASTER CERES EOS-COLOR EOSP MISR MODIS MOPITT	Sources/sinks of CO_2, CH_4, and NOx as relate to land-surface cover Radiation budget and multi-decadal studies to differentiate forcings Ocean productivity to measure ocean's capacity to absorb atmospheric CO_2 The indirect effect of aerosols as cloud condensation nuclei Bidirectional reflectance to distinguish human-induced land surface changes Carbon balance of terrestrial and oceanic ecosystems under different climate and/or land management conditions CO_2 and CH_4 as relate to human-induced global warming potential
	What are the uncertainties and potential for surprises, including sudden changes in the frequency and intensity of extreme events?	AIRS GLAS MODIS SAGE III SeaWinds	Role of oceans in global heat transport First opportunity to determine if the major ice sheets are growing or shrinking Understand uncertainties surrounding cloud-climate feedbacks Volcanic eruptions, which can affect climate for years Addressing uncertainties of air-sea interactions

ecosystems, with the affected species' only recourse to adapt quickly or die. Land management practices introduce patchiness or large-scale changes in habitats and modify ecosystem function, including extinction, migration pathways, and survival. The Landsat Pathfinder study indicated that the effect of deforestation on biodiversity in the Amazon is greater than previously thought (Skole and Tucker, 1993). The effect on regional hydrology and climatology could prove even more significant. A lack of well-documented and quantitative global data sets limits a clear understanding of both local and large-scale impacts on Earth system processes.

EOS instruments will map changes in land-use patterns, vegetation type, and photosynthetic capacity of ecosystems for supporting habitats (see Table 9). The EOS Program is sponsoring extensive sensitivity experiments on climate models at regional to global scales. These climate predictions are a prerequisite for assessing uncertainties and the potential for surprises associated with ecosystem response to climate change. Better climate change predictions will contribute to investigations of ecosystem response in both terrestrial and oceanic realms, based on a better understanding of carbon and nitrogen balances, ozone depletion, deforestation, ocean circulation, air-sea interaction, and other climate forcing factors. In addition, EOS will monitor volcanic eruptions as a natural disruptive force to ecosystems.

The economic tradeoffs associated with the use of both natural and managed ecosystems for agricultural and industrial development must be based on sound scientific understanding. The EOS Program will provide the knowledge required

TABLE 9. Ecosystems and Biodiversity

global change ISSUE	priority research QUESTIONS	relevant EOS INSTRUMENTS	key CONTRIBUTIONS
Ecosystems and Biodiversity	What have been and will be the impacts of global change on terrestrial systems and their wildlife—including forests, grasslands, arid lands, the tropics, and high latitudes?	ASTER MISR MLS MODIS MOPITT SAGE III	Changes in land-cover patterns—including expansion/contraction of farmland, urban growth, deforestation, and forest regrowth—related to land use or environmental change Changes in grasslands, tundra, forests, and tropical deforestation to serve as basis for ecosystem response studies Monitoring of ozone and greenhouse forcing, and their impact on ecosystems and potential biome shifts Vegetation cover and habitat type/extent, and impact on biodiversity Trends related to increased atmospheric CO_2, and its impact on the carbon pool (e.g., increased sequestration of CO_2 through forest management practices) The role of volcanism on climate and its impact on atmospheric temperature, radiation, etc.
	What have been and will be the impacts of global change on coastal and marine ecosystems?	GLAS EOS-COLOR MODIS SeaWinds SSALT	Mass balance of glaciers and ice sheets to assess possible sea-level change Shifts in ocean productivity and impact on marine life Impact on coastal and marine ecosystems Wind stress forcings in ocean circulation and productivity Ecosystem response to air-sea fluxes which influence ocean circulation
	What are the uncertainties and the potential for surprises and unexpected changes in the global ecosystem?	ASTER MISR MODIS	Sensitivity of ecosystems to global change to be achieved through these and other EOS instruments via data-constrained simulation models of carbon and nitrogen balance Extensive climate model sensitivity experiments for the full host of forcing factors at regional to global scales Assessment of uncertainties in ecosystem response to climate change

by exploiting existing resources and by providing a conduit to the vast capabilities of IEOS, as-yet-undefined future national and international missions, and the consequent international data and information holdings.

Social and Economic Implications

The development of a predictive understanding of how human activities affect and are affected by the Earth system is a complex task. Human influences and interactions with the global environment cover spatial scales from local to global, and extend over temporal scales ranging from minutes to centuries. All segments of human enterprise participate in this global experiment, both on individual and collective bases. As such, global change research must further understanding of social and institutional systems, as well as their physical, chemical, and biological counterparts. The effect on human society needs to be gauged, since each of us is intimately tied to the world about us. Climate change affects all, and all are interested in the character and quality of global change predictions. The EOS Program seeks to better define the relationship and response of human society to global change, and to help develop rational adaptation and mitigation strategies for consideration by policymakers around the world.

Knowledge of the rate and magnitude of global change and its effect at global and regional scales proves essential. Strategies for mitigation and adaptation may have widely different socio-economic consequences, involving health, standard of living, and quality of life. In addition, government officials need to consider the likelihood of implementation given the redistribution of resources that global change could promote. Behavior—including valuation of the environment, issues of equity among generations, global change decisionmaking, and the role of institutions—becomes an essential element for assessing the human dimension of global change. Without these foundations, the strategies for mitigation or adaptation may be misguided, resulting in either inadequate or unnecessary policies.

These overarching issues fall within the realm of the U.S. Global Change Research Program and the larger international oversight bodies (e.g., UNEP and WMO). However, EOS provides the foundation for their socio-economic evaluations, based on the physical manifestations of global climate change. EOS examines natural hazards and provides a realistic basis for understanding the potential rate and magnitude of global change. In addition, the EOS Program contributes to understanding of perhaps the most significant human forcing factor—land-cover change. Table 10 provides examples of EOS contributions to issues such as water resource vulnerability, agriculture, and ecosystems' response, which can directly affect land management practices, business and commerce, and human health assessments.

TABLE 10. Social and Economic Implications of Global Climate Change

global change ISSUE	priority research QUESTIONS	relevant EOS INSTRUMENTS	key CONTRIBUTIONS
Social and Economic Implications	What are the interactions with land cover?	ALL EOS INSTRUMENTS	Observed changes in land cover, surface albedo, etc., as basis to assess human activity and to define climate-agriculture interactions
	Freshwater resources?		Emphasis on water balance at all scales to assess water resource availability/vulnerability
	Coastal zones, including fisheries?		High-resolution coastal circulation studies (including biology) to provide important data on primary productivity, a key element in fisheries
	Human health, including disease vectors, air quality, and UV-B radiation?		Advances in understanding of tropospheric pollution and stratospheric ozone depletion (thereby anticipating UV-B increases) Identification of volcanic gas hazards Early-warning capability for natural disasters Ability to isolate as-yet-undiscovered risks to human health and safety
	Land use, including soils and erosion?		Unprecedented measurement of the scale and magnitude of land cover associated with land-use management and practices
	Business and commerce?		Identification of agricultural implications such as water resource availability Identification of volcanic hazards (e.g., disruption of commercial air traffic) Affect on regional commerce Management of renewable and non-renewable natural resources

ASSESSMENT OF CLIMATE CHANGE

The EOS series of satellites, associated instrument complements, international Earth science data and information holdings, and interdisciplinary teams of investigators and modelers will generate a wealth of information to support global change research and policy. EOS instrument capabilities are vast, with a highly talented pool of scientists already on hand and still being trained through educational programs and grants awarded to U.S. academic institutions. Yet, gifted scientists and a bevy of instrument observations accomplish little if findings cannot be communicated to the policymakers and public in a timely and easily decipherable manner. Calibrated instrument radiance measurements do not automatically translate into sound environmental policies.

Problems can be identified, but successful implementation of proposed mitigation measures takes years of negotiation and decades of validation, taking into account the individual biases of the nations involved and the limited capability to enforce international decrees. Just as we are uncertain of the response of the Earth to increased greenhouse gas emissions, we are uncertain how people and institutions will react to changes in the natural and human environments. Future patterns of conduct do not readily conform to past policy actions, and global change research is not the forum to police international agreements. Rather, sustained consensus-building must bridge the gap. Understanding the interaction of human and environmental systems and developing realistic global change mitigation scenarios have both led to an increased emphasis on programs that explore policies and choices, especially in the areas of integrated assessments and socio-economic sciences.

Integrated assessments involve the synthesis of relevant biophysical and socio-economic research on environmental change, its causes, and its effects. This process brings research results from natural, social, and policy sciences into a framework that helps decisionmakers formulate and evaluate actions to respond to specific environmental changes. Though focusing on a single issue, this mechanism attempts to provide information about the secondary effects for each of the potential policy actions. This capability is still primitive.

EOS contributes to the refinement of Integrated Assessment Models (IAMs) by pioneering the synthesis of disciplinary models, which had formerly been constructed with different technical approaches, characteristics, and parameterization schemes. Ultimately, IAMs will link human forcing functions (e.g., greenhouse gas emissions), the effect on the physical Earth system, effects on humans and ecosystems, and the economic and environmental consequences of various potential policy actions. To be effective, these models must accurately reflect the fundamental relationships between policy goals and the social processes required to realize them. Response planning requires a predictive understanding of the various ways that individuals, institutions, and societies react to environmental changes and the policies or programs intended to mitigate the consequences of global change.

The current concern over ozone depletion, climate change, and the reduction of biodiversity has resulted in an international endorsement that comprehensive scientific environmental impact and technical and economic assessments be undertaken now. IAMs will draw heavily on the modeling breakthroughs anticipated from EOS interdisciplinary investigations. As evidenced by the Montreal Protocol and the Climate Convention Protocol, international assessments have already had a profound impact on global policy formulation. Advances in numerical modeling can only enhance this process. Though EOS hardware has yet to be launched, EOS-funded investigators are already making significant contributions to the international assessments described below:

- In response to a mandate from the Vienna Convention for Protection of the Ozone Layer and the Montreal Protocol on Substances that Deplete the Ozone Layer, WMO and UNEP are sponsoring comprehensive assessments on ozone. To be completed by late 1994, this latest state-of-the-science assessment addresses the following: 1) The natural processes and human activities controlling the present distribution and rate of change of atmospheric ozone; 2) the impacts of ozone depletion on human health, and managed and natural ecosystems; and 3) the technological feasibility and economic effects associated with the substitution of substances controlled under the Montreal Protocol.

- Under the auspices of WMO and UNEP, the IPCC has undertaken comprehensive climate assessments. The First Assessment Report was released in 1990, with a supplement in 1992. The Second Assessment Report, slated to be released in mid-1995, addresses: 1) The entire spectrum of climate change issues, such as climate forcing, tests of climate models, past climate records, simulations of past climatic changes, predictions of future climate, and biological and physical responses to climate change; 2) current knowledge of the impacts of and adaptation to climate change on human health, natural ecological systems, and socioeconomic sectors (e.g., agriculture, forestry, fisheries, water resources, industrial sectors, and human settlements); and 3) the economic aspects of climate change. Two interim assessment activities are also underway; they will be completed by mid-1994, before the first meeting of the Climate Convention. These reports will assess the relative radiative forcing of different greenhouse gases, establish a methodology for calculating net greenhouse gas inventories at the national level, and examine the validity of the greenhouse gas emissions scenarios developed for the 1992 IPCC assessment.

- Under the auspices of UNEP and at the request of the Biological Diversity Convention, a comprehensive global biodiversity assessment has recently been undertaken. To be completed in 1995, this assessment seeks to 1) characterize biodiversity; 2) understand its origin, dynamics, and future; 3) assess its magnitude and distribution; 4) understand the basic principles of biodiversity and ecosystem function; 5) discern human influences; 6) determine conservation, restoration, sustainable use, and maintenance methods; and 7) strengthen data and information management and communication.

As part of its distributed data system, NASA has directed CIESIN to develop and operate a Socio-Economic Data and Applications Center (SEDAC). The other eight EOS DAACs archive natural science data, while SEDAC focuses on human interactions in global environmental change. Human activities heavily influence the Earth's physical and biological systems, so further study of the human dimensions associated with global change proves integral if researchers are to accurately anticipate global change and make informed recommendations about the human response. To assess human feedback to the Earth system and assist in the IAMs mentioned above, CIESIN has established an Information Cooperative to help identify, collect, and make accessible relevant socio-economic and environmental data. An international conglomeration of institutions has agreed to catalog and share their information electronically, performing a role within SEDAC that the EOS satellites will play for the other DAACs. This cooperative serves as a platform for observing and recording economic and policy data; population, settlement, and transportation trends; and the dynamics of anthropogenic emissions and pollution.

The EOS Program already provides the scientific and technical expertise to aid in integrated assessments. Access to global change data has been made easier with the implementation of EOSDIS Version 0, and all interested parties can secure Earth science data if they meet the liberal qualification standards. This enlightened data policy ensures fair and equitable access to all, and can only enhance the quality of research expected from the EOS Program.

CONCLUSIONS

We think of exploration as a process of humanity leaving planet Earth to go out into the solar system and beyond. However, from the orbiting of the first satellite to the flights of the first humans into space, there has also been a desire to look back, to see the Earth from space and understand it in a new way. This process has led to Mission to Planet Earth, a worldwide enterprise aimed at understanding our home planet with a depth and breadth that we could not have imagined even a few decades ago.

The many satellites, instruments, interdisciplinary studies, information systems, variables, and unknowns involved in the effort make the strategy behind it difficult to comprehend; yet, the vision that inspires the mission is easy to grasp. It is the vision of humanity taking greater responsibility for the Earth system, basing policy decisions about the environment on scientific understanding of global change. It is the vision of humanity predicting the effects of its actions on the natural system, thereby adapting to new conditions and/or ameliorating negative impacts.

The question of whether or not human society induces climate change is no longer in doubt. Therefore, we must meld comprehensive hindsight to understand the repercussions associated with industrial advancement with judicious foresight to anticipate plausible global implications. All segments of the user community must help augment the capabilities of EOS and those of our international partners. EOS

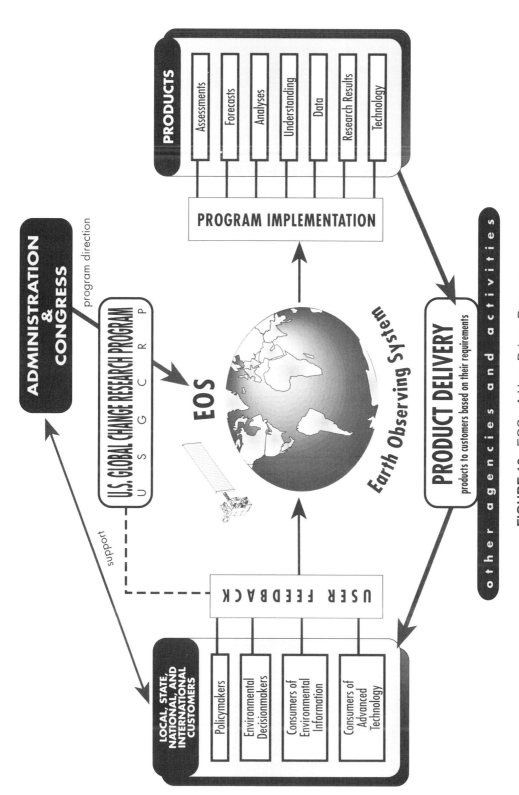

FIGURE 10. EOS—A User-Driven Program

actively solicits information and opinions from everyone involved in the global change arena because of the need to consider countless factors—from individual quality of life to global economic impact—when developing sound environmental policy. The EOS Program is flexible enough to respond readily to changing user requirements, evolving as our understanding of the Earth system and global change becomes more refined.

EOS provides the means to assess humanity's impact on the Earth system. It is an opportunity for those interested in space exploration and those interested in preserving the environment to come together in a common cause—to use the view of the Earth from space to protect the planet and all living systems on it.

APPENDIX A

Acronyms

AATSR	Advanced Along-Track Scanning Radiometer
ACRIM	Active Cavity Radiometer Irradiance Monitor
ADALT	Advanced Radar Altimeter
ADC	Affiliated Data Center
ADEOS	Advanced Earth Observing System
AIRS	Atmospheric Infrared Sounder
AMSR	Advanced Microwave Scanning Radiometer
AMSU	Advanced Microwave Sounding Unit
AO	Announcement of Opportunity
ARM	Atmospheric Radiation Measurement
ASAR	Advanced Synthetic Aperture Radar
ASF	Alaska SAR Facility
ASCAT	Advanced Scatterometer
ASTER	Advanced Spaceborne Thermal Emission and Reflection Radiometer
ATLAS	Atmospheric Laboratory for Applications and Science
AVHRR	Advanced Very High-Resolution Radiometer
AVNIR	Advanced Visible and Near-Infrared Radiometer
BOREAS	Boreal Ecosystem-Atmosphere Study
C	Centigrade
CEES	Committee on Earth and Environmental Sciences
CENR	Committee on the Environment and Natural Resources
CERES	Clouds and Earth's Radiant Energy System
CFC	Chlorofluorocarbon
CIESIN	Consortium for International Earth Science Information Network
CNES	Centre National d'Etudes Spatiales
CRYSYS	Cryospheric System
CSIRO	Commonwealth Scientific and Industrial Research Organization
CZCS	Coastal Zone Color Scanner
DAAC	Distributed Active Archive Center
DADS	Data Archive and Distribution System
DCS	Data Collection System
DIAL	Differential Absorption Lidar
DMSP	Defense Meteorological Satellite Program
DOC	Dissolved Organic Carbon
DORIS	Doppler Orbitography and Radiopositioning Integrated by Satellite
DRSS	Data Relay Satellite System

EDC	EROS Data Center
EDOS	EOS Data and Operations System
E-LIDAR	Experimental Light Detection and Ranging
ENSO	El Niño-Southern Oscillation
ENVISAT	Environmental Satellite
EOC	EOS Operations Center
EOS	Earth Observing System
EOSDIS	EOS Data and Information System
EOSP	Earth Observing Scanning Polarimeter
EP	Earth Probe
EPA	Environmental Protection Agency
ERBE	Earth Radiation Budget Experiment
EROS	Earth Resources Observation System
ERS	European Remote-Sensing Satellite
ESA	European Space Agency
FIRE	First ISCCP Regional Experiment
GEWEX	Global Energy and Water Cycle Experiment
GLAS	Geoscience Laser Altimeter System
GLI	Global Imager
GMS	Geostationary Meteorological Satellite
GOES	Geostationary Operational Environmental Satellite
GOMI	Global Ozone Monitoring Instrument
GOMOS	Global Ozone Monitoring by Occultation of Stars
GPS	Global Positioning System
GSFC	Goddard Space Flight Center
HDF	Hierarchical Data Format
HIRDLS	High-Resolution Dynamics Limb Sounder
HIROS	High-Resolution Observation System
HIRS	High-Resolution Infrared Sounder
IAM	Integrated Assessment Model
IASI	Infrared Atmospheric Sounding Interferometer
IDS	Interdisciplinary Science Investigation
IGBP	International Geosphere-Biosphere Program
IELV	Intermediate Expendable Launch Vehicle
IEOS	International Earth Observing System
ILAS	Improved Limb Atmospheric Spectrometer
IMB	Investigator of Micro-Biosphere
IMG	Interferometric Monitor for Greenhouse Gases
IMS	Information Management System
IPCC	Intergovernmental Panel on Climate Change
IPOC	International Partner Operations Center
IR	Infrared
ISCCP	International Satellite Cloud Climatology Project
ISLSCP	International Satellite Land Surface Climatology Project
IST	Instrument Support Terminal
JERS	Japan's Earth Resources Satellite
JGOFS	Joint Global Ocean Flux Study
JPL	Jet Propulsion Laboratory
K	Kelvin
Landsat	Land Remote-Sensing Satellite
LaRC	Langley Research Center

LERTS	Laboratoire d'Etudes et de Recherches en Teledetection Spatiale
LIS	Lightning Imaging Sensor
LITE	Lidar In-Space Technology Experiment
MAPS	Measurement of Air Pollution from Satellites
MCP	Meteorological Communications Package
MELV	Medium Expendable Launch Vehicle
MERIS	Medium-Resolution Imaging Spectrometer
Meteosat	Meteorological Satellite
METOP	Meteorological Operational Satellite
MHS	Microwave Humidity Sounder
MIMR	Multifrequency Imaging Microwave Radiometer
MIPAS	Michelson Interferometer for Passive Atmospheric Sounding
MISR	Multi-Angle Imaging SpectroRadiometer
MLS	Microwave Limb Sounder
MODIS	Moderate-Resolution Imaging Spectroradiometer
MOPITT	Measurements of Pollution in the Troposphere
MOS	Marine Observation Satellite
MSFC	Marshall Space Flight Center
MTPE	Mission to Planet Earth
MWR	Microwave Radiometer
NASA	National Aeronautics and Space Administration
NASDA	National Space Development Agency
NCAR	National Center for Atmospheric Research
NDSC	Network for Detection of Stratospheric Change
NOAA	National Oceanic and Atmospheric Administration
NREN	National Research and Education Network
NSCAT	NASA Scatterometer
NSIDC	National Snow and Ice Data Center
NSTC	National Science and Technology Council
OCTS	Ocean Color and Temperature Scanner
ORNL	Oak Ridge National Laboratory
PGS	Product Generation System
PI	Principal Investigator
POES	Polar-Orbiting Operational Environmental Satellite
POLDER	Polarization and Directionality of Earth's Reflectances
POLES	Polar Exchange at the Sea Surface
PR	Precipitation Radar
QC	Quality Control
RA	Radar Altimeter
Radarsat	Radar Satellite
RIS	Retroreflector in Space
S&R	Search and Rescue
SAGE	Stratospheric Aerosol and Gas Experiment
SAR	Synthetic Aperture Radar
SBUV	Solar Backscatter Ultraviolet
SCARAB	Scanner for the Radiation Budget
SCF	Science Computing Facility
SCIAMACHY	Scanning Imaging Absorption Spectrometer for Atmospheric Cartography
SeaWiFS	Sea-Viewing Wide Field Sensor
SEDAC	Socio-Economic Data and Applications Center
SELV	Small Expendable Launch Vehicle

SEM	Space Environment Monitor
SIR-C	Shuttle Imaging Radar-C
SLIES	Stratospheric Limb Infrared Emission Spectrometer
SMC	System Management Center
SMM	Solar Maximum Mission
SMMR	Scanning Multispectral Microwave Radiometer
SOHO	Solar and Heliospheric Observatory
SOLSTICE	Solar Stellar Irradiance Comparison Experiment
SPOT	Systeme pour l'Observation de la Terre
SSALT	Solid-State Altimeter
SSBUV	Shuttle Solar Backscatter Ultraviolet
SSM/I	Special Sensor Microwave/Imager
TDRSS	Tracking and Data Relay Satellite System
TERSE	Tunable Etalon Remote Sounder of Earth
TES	Tropospheric Emission Spectrometer
TIROS	Television Infrared Observing Satellite
TMI	TRMM Microwave Imager
TMR	TOPEX Microwave Radiometer
TOMS	Total Ozone Mapping Spectrometer
TOMUIS	3-D Ozone Mapping with Ultraviolet Imaging Spectrometer
TOPEX	Ocean Topography Experiment
TOVS	TIROS Operational Vertical Sounder
TRMM	Tropical Rainfall Measuring Mission
UARS	Upper Atmosphere Research Satellite
UNCED	United Nations Conference on Environment and Development
UNEP	United Nations Environment Program
USGCRP	U.S. Global Change Research Program
UV-B	Ultraviolet-B
VIRS	Visible Infrared Scanner
WCRP	World Climate Research Program
WMO	World Meteorological Organization
WOCE	World Ocean Circulation Experiment
X-SAR	X-Band Synthetic Aperture Radar

APPENDIX B

Bibliography

Abbott, M.R. and M.H. Freilich, 1992: Report of the EOS Oceans Panel to the Payload Panel. *Palaeogeography, Palaeoclimatology, Palaeoecology*, 98, pp. 25–28.

Andreae, M.O. and D.S. Schimel (eds.), 1989: *Exchange of Trace Gases between Terrestrial Ecosystems and the Atmosphere*, John Wiley and Sons, New York, New York, 346 pp.

Asrar, G., S.G. Tilford, and D.M. Butler, 1992: Mission to Planet Earth: Earth Observing System (EOS). *Global and Planetary Change*, 6, Amsterdam, Netherlands, 18 pp.

Backlund, P., L. Shaffer, and D. Dokken, 1989: *Mission to Planet Earth*, National Aeronautics and Space Administration, Washington, D.C., 71 pp.

Baker, D.J., 1990: *Planet Earth, The View From Space*, Harvard University Press, Cambridge, Massachusetts, 145 pp.

Ballish, B., X. Cao, E. Kalnay, and M. Kanamitsu, 1992: Incremental nonlinear normal-mode initialization. *Monthly Weather Review*, 120, pp. 1723–1734.

Barry, R.G., J. Maslanik, K. Steffen, R.L. Weaver, V. Troisi, D.J. Cavalieri, and S. Martin, 1993: Advances in sea ice research based on remotely sensed passive microwave data. *Oceanography*, 6, pp. 5-13.

Bates, J.R., S. Moorthi, and R.W. Higgins, 1993: A global multilevel atmospheric model using a vector semi-Lagrangian finite difference scheme. Part I: Dynamics. *Monthly Weather Review*, 121, pp. 244–263.

Bell, T.L. and N. Reid, 1993: Detecting the diurnal cycle of rainfall using satellite observations. *Journal of Applied Meteorology*, 32, pp. 311–322.

Brutsaert, W., A. Hsu, and T.J. Schmugge, 1993: Parameterization of surface heat fluxes above forest with satellite thermal sensing and boundary layer soundings. *Journal of Applied Meteorology*, 32, pp. 909–917.

Brutsaert, W. and M.B. Parlange, 1992: The unstable surface layer above forest: Regional evaporation and heat flux. *Water Resources Research*, 28, pp. 3129–3134.

Carson, R., 1962: *Silent Spring.* Houghton Mifflin, Boston, 368 pp.

Chahine, M.T., 1992: The hydrological cycle and its influence on climate. *Nature,* 359, pp. 373–380.

Chen, T.-C. and J. Pfaendtner, 1993: On the atmospheric branch of the hydrological cycle. *Journal of Climate,* 6, p. 161.

Committee on Earth and Environmental Sciences, 1992: *Our Changing Planet: The FY 1993 U.S. Global Change Research Program,* National Science Foundation, Washington, D.C., 79 pp.

Committee on Earth and Environmental Sciences, 1991: *Our Changing Planet: The FY 1992 U.S. Global Change Research Program,* U.S. Geological Survey, Reston, Virginia, 90 pp.

Committee on Earth and Environmental Sciences, 1990: *Our Changing Planet: The FY 1991 U.S. Global Change Research Program,* U.S. Geological Survey, Reston, Virginia, 90 pp.

Committee on Earth and Environmental Sciences, 1990: *Our Changing Planet: The FY91 Research Plan,* U.S. Geological Survey, Reston, Virginia, 260 pp

Crutzen, P.J. and P.H. Zimmerman, 1991: The changing photochemistry of the troposphere. *Tellus,* AB, pp. 136–151.

Crutzen, P.J. and M.O. Andreae, 1990: Biomass burning in the tropics: Impacts on atmospheric chemistry and biogeochemical cycles. *Science,* 250, pp. 1169–1678.

DelGenio, A.D., 1993: Convective and large-scale cloud processes in global climate models. In *Energy and Water Cycles in the Climate System.* Proceedings of the NATO Advanced Study Institute, September 30 - October 11, 1991, Glucksburg, Germany.

Dickinson, R.E. and R.J. Cicerone, 1986: Future global warming from atmospheric trace gases. *Nature,* 319, pp. 109–115.

Diner, D.J., C.J. Bruegge, J.V. Martonchik, G.W. Bothwell, E.D. Danielson, E.L. Floyd, V.G. Ford, L.E. Hovland, K.L. Jones, and M.L. White, 1991: A Multiangle Imaging SpectroRadiometer for terrestrial remote sensing from the Earth Observing System. *International Journal of Imaging Systems and Technology,* 3, pp. 92–107.

Dozier, J. and H.K. Ramapriyan, 1991: Planning for the EOS Data and Information System (EOS-DIS). In *Global Environmental Change,* NATO ASI Series I: Global Environmental Change, Vol. 1. Edited by R.W. Correll and P.A. Anderson. Springer-Verlag, Berlin, pp. 155–180.

Esbensen, S.K., D.B. Chelton, D. Vickers, and J. Sun, 1993: An analysis of errors in Special Sensor Microwave Imager evaporation estimates over the global oceans. *Journal of Geophysical Research,* 98C, pp. 7081–7101.

Fasham, M.J.R., H.W. Ducklow, and S.M. McKelvie, 1990: A nitrogen-based model of plankton dynamics in the oceanic mixed layer. *Journal of Marine Research,* 48, pp. 591–639.

Field, C.B., 1992: Ecological scaling of carbon gain to stress and resource availability. In *Integrated Responses of Plants to Stress.* Edited by H.A. Mooney, W.E. Winner, and E.J. Pell. Academic Press.

Fox, A.N. and M.A. Strecker, 1991: Pleistocene and modern snowlines in the Central Andes (24-28°S). *Bamberger Geographische Schriften*, 11, pp. 169-182.

Gao, X., S. Sorooshian, and D.C. Goodrich, 1993: Linkage of a Geographic Information System (GIS) to a distributed rainfall-runoff model. In *Environmental Modeling with GIS*. Edited by M.F. Goodchild et al. Oxford University Press, pp. 182–187.

Goodison, B.E. and A.E. Walker, 1993: Canadian development and use of snow cover information from passive microwave satellite data. *Proceedings ESA/NASA International Workshop on Passive Remote Sensing Research Related to Land-Atmosphere Interactions*, St. Lary, France, January 11–15, 1993.

Gupta, S.K., W.L. Darnell, and A.C. Wilber, 1992: A parameterization for surface longwave radiation from satellite data: Recent improvements. *Journal of Applied Meteorology*, 31, pp. 1361–1367.

Gurney, R.J., J.L. Foster, C.L. Parkinson (eds.), 1993: *Atlas of Satellite Observations Related to Global Change*. Cambridge University Press, 470 pp.

Hansen, J., A. Lacis, R. Ruedy, M. Sato, and H. Wilson, 1993: How sensitive is the world's climate? *National Geographic Research and Exploration*, 9, pp. 142–158.

Hansen, J., I. Fung, A. Lacis, D. Rind, S. Lebedeff, R. Ruedy, G. Russell, and P. Stone, 1988: Global climate changes as forecast by the Goddard Institute for Space Studies three-dimensional model. *Journal of Geophysical Research*, 93, pp. 9341–9364.

Harrison, E.F., P. Minnis, G.G. Gibson, and F.M. Denn, 1991: Orbital analysis and instrument viewing considerations for the Earth Observing System (EOS) satellite. *Advances in Astronautical Sciences*, 76, pp. 1215–1228.

Hook, S.J., A.R. Gabell, A.A. Green, and P.S. Kealy, 1992: A comparison of techniques for extracting emissivity information from thermal infrared data for geologic studies. *Remote Sensing of Environment*, 42, pp. 123–135.

Huete, A.R., G. Hua, J. Qi, A. Chehbouni, and W.J.D. van Leeuwen, 1992: Normalization of multidirectional red and NIR reflectances with the SAVI. *Remote Sensing of Environment*, 41, pp. 143–154.

Intergovernmental Panel on Climate Change, 1990: *Climate Change: The IPCC Scientific Assessment*. Edited by J.T. Houghton, G.J. Jenkins, and J.J. Ephraums. Cambridge University Press, Cambridge, 365 pp.

Isacks, B.L. and P. Mouginis-Mark, 1992: Solid Earth science in EOS—Report of the Solid Earth Panel. *Global and Planetary Change*, 6, pp. 29–35.

Jenne, R.L., 1992: Climate model description and impact on terrestrial climate. In *Global Climate Change: Implications, Challenges, and Mitigation Measures*. Edited by S.K. Majumdar et al. The Pennsylvania Acadamy of Science, pp. 145–164.

Kaufman, Y.J. and M.-D. Chou, 1993: Model simulations of the competing climatic effects of SO_2 and CO_2. *Journal of Climate*, 6, pp. 1241–1252.

Kaufman, Y.J. and B.-C. Gao, 1992: Remote sensing of water vapor in the near IR from EOS/MODIS. *IEEE Transactions on Geoscience and Remote Sensing*, 30, pp. 871–888.

Kelly, P.M. and T.M.L. Wigley, 1992: Solar cycle length, greenhouse forcing, and global climate. *Nature*, 360, pp. 328–330b.

Key, J. and M. Haefliger, 1992: Arctic ice surface temperature retrieval from AVHRR thermal channels. *Journal of Geophysical Research*, 97D, pp. 5885–5893.

King, M.D., 1992: Remote sensing of cloud, aerosol, and water vapor properties from the Moderate-Resolution Imaging Spectrometer (MODIS). In *The Use of EOS for Studies of Atmospheric Physics*. Editrice Compositori, Bologna, Italy.

Klein, S.A. and D.L. Hartmann, 1993: Spurious changes in the ISCCP data set. *Geophysical Research Letters*, 20, pp. 455–458.

Koblinsky, C.J., P. Gaspar, and G. Lagerloef (eds.), 1992: *The Future of Spaceborne Altimetry: Oceans and Climate Change*. Joint Oceanographic Institutions, Inc., Washington, D.C. 75 pp.

Lacis, A.A., D.J. Wuebbles, and J.A. Logan, 1990: Radiative forcing of climate by changes of the vertical distribution of ozone. *Journal of Geophysical Research*, 95, pp. 9971–9981.

Leslie, L.M. and K. Fraedrich, 1990: Reduction of tropical cyclone position errors using an optimal combination of independent forecasts. *Weather Forecasting*, 5, pp. 158–161.

Liu, W.T., 1992: Spaceborne scatterometer in synergistic studies of global changes. *Remote Sensing of Environment*, 93 pp. 284–293.

Logan, J.A., 1983: Nitrogen oxides in the atmosphere: Global and regional budgets. *Journal of Geophysical Research*, 88, pp. 10785–10936.

Martonchik, J.V. and D.J. Diner, 1992: Retrieval of aerosol optical properties from multi-angle satellite imagery. *IEEE Transactions on Geoscience and Remote Sensing*, 30, pp. 223–230.

Matson, P.A., P.M. Vitousek, G.P. Livingston, and N.A. Swanberg, 1990: Sources of variation in nitrous oxide flux from Amazonian ecosystems. *Journal of Geophysical Research*, 95-16, 789–16, 798.

Mertes, L.A.K., M. Smith, and J.B. Adams, 1993: Quantifying sediment concentration on surface waters of the Amazon River wetlands from Landsat images. *Remote Sensing of Environment*, 43, pp. 281–301.

Moore, B., J. Dozier, M.R. Abbott, D.M. Butler, D. Schimel, and M.R. Schoerberl, 1991: The restructured Earth Observing System: Instrument recommendations. *EOS Transactions*, 72(46), pp. 505–516.

Nakazawa, T., 1992: Seasonal phase lock of intraseasonal variation during the Asian summer monsoon. *Journal of the Meteorological Society of Japan*, 70, pp. 597–611.

National Aeronautics and Space Administration, 1993: *NASA's Mission to Planet Earth: Catalog of Education Programs and Activities*. Office of Mission to Planet Earth and Office of Human Resources and Education, NP-206, 40 pp.

National Aeronautics and Space Administration, 1989: *The Upper Atmosphere Research Satellite (UARS): A Program to Study Global Ozone Change*, National Aeronautics and Space Administration, Washington, D.C., 28 pp.

National Aeronautics and Space Administration, 1987: *Arctic Sea Ice, 1973-1976: Satellite Passive-Microwave Observations*. GSFC Scientific and Technical Information Branch, NASA SP-489, 296 pp.

National Aeronautics and Space Administration, 1987: *From Pattern to Process: The Strategy of the Earth Observing System*, EOS Science Steering Committee Report, Vol. II, Goddard Printing Office, Greenbelt, Maryland, 140 pp.

Nemani, R., L. Pierce, and S. Running, 1993: Developing satellite-derived estimates of surface moisture status. *Journal of Applied Meteorology*, 32, pp. 548–557.

Nerem, R.S., B.D. Tapley, and C.K. Shum, 1990: Determination of ocean circulation using Geosat altimetry. *Journal of Geophysical Research*, 95, pp. 3163–3179.

Nolin, A.W. and J. Dozier, 1993: Estimating snow grain size using AVIRIS data. *Remote Sensing of the Environment*, 44, pp. 231–238.

Office of Science and Technology Policy, 1991: *Policy Statements on Data Management for Global Change Research*, National Science Foundation, Washington, D.C., DOE/EP-0001P, 6 pp.

Perry, J.S., 1992: *The U.S. Global Change Research Program: Early Achievements and Future Directions*, prepared for the National Research Council Board on Global Change, National Academy Press, Washington, D.C., 20 pp.

Ponte, R.M., D.A. Salstein, and R.D. Rosen, 1993: Determining torques over the ocean and their role in the planetary momentum budget. *Journal of Geophysical Research*, 98D, pp. 7317–7325.

Ponte, R.M., D.A. Salstein, and R.D. Rosen, 1991: Sea level response to pressure forcing in a barotropic numerical model. *Journal of Physical Oceanography*, 21, pp. 1043–1057.

Puri, K.K., N.E. Davidson, L.M. Leslie, and L.W. Logan, 1992: The BMRC tropical limited-area model. *Australian Meteorology Magazine*, 40, pp. 81–104.

Ramanathan, V., B.R. Barkstrom, and E.F. Harrison, 1989: Clouds and the Earth's radiation budget. *Physics Today*, 42, pp. 22–32.

Rampino, M.R., S. Self, and R.B. Stothers, 1988: Volcanic winters. *Annual Review of Earth Planetary Science*, 16, pp. 73–99.

Rind, D., R. Goldberg, J. Hansen, C. Rosenzweig, and R. Ruedy, 1990: Potential evapotranspiration and the likelihood of future drought. *Journal of Geophysical Research*, 95, pp. 983–1004.

Rood, R., A. Douglass, and C. Weaver, 1992: Tracer exchange between tropics and middle latitudes. *Geophysical Research Letters*, 19, pp. 805–808.

Rossow, W.B. and R.A. Schiffer, 1991: ISCCP cloud data products. *Bulletin of the American Meteorological Society*, 72, pp. 2–20.

Sabol, D.E., J.B. Adams, and M.O. Smith, 1992: Quantitative sub-pixel spectral detection of targets in multispectral images. *Journal of Geophysical Research*, 97, pp. 2659–2672.

Schoeberl, M., J. Pfaendtner, R. Rood, A. Thompson, and B. Wielicki, 1992: Atmospheres Panel Report to the Payload Panel. *Palaeogeography, Palaeoclimatology, Palaeoecology*, 98, pp. 9–21.

Schubert, S., R. Rood, and J. Pfaendtner, 1993: An assimilated data set for Earth science applications. *Bulletin of the American Meteorological Society*, 74, pp. 2331–2342.

Schweiger, A.J., M.C. Serreze, and J. Key, 1993: Arctic sea ice albedo: A comparison of two satellite-derived data sets. *Geophysical Research Letters*, 20, pp. 41-44.

Sellers, P.J. and D.S. Schimel, 1993: Remote sensing of the land biosphere and biogeochemistry in the EOS era: Science priorities, methods, and implementation—EOS Land Biosphere and Biogeochemical Cycles Panels. *Global and Planetary Change*, pp. 279–297.

Shukla, J., C. Nobre, and P.J. Sellers, 1990: Amazon deforestation and climate change. *Science*, 247, pp. 1322–1325.

Simkin, T., 1993: Terrestrial volcanism in space and time. *Annual Review of Earth Planetary Science*, 21, pp. 427–452.

Sippel, S.J., S.K. Hamilton, and J.M. Melack, 1992: Inundation area and morphometry of lakes on the Amazon River floodplain, Brazil. *Archiv fur Hydrobiologie*, 123, pp. 385–400.

Skole, D. and C. Tucker, 1993: Tropical deforestation and habitat fragmentation in the Amazon: Satellite data from 1978 to 1988. *Science*, 260, pp. 1849–2024.

Takayabu, Y.N. and M. Murakami, 1991: The structure of supercloud clusters observed in 1-20 June 1986 and the relationship to easterly waves. *Journal of the Meteorological Society of Japan*, 69, pp. 105–125.

Tans, P.P., I.Y. Fung, and T. Takahashi, 1990: Observational constraints on the global atmospheric carbon dioxide budget. *Science*, 247, pp. 1431–1438.

Teillet, P.M. and R.P. Santer, 1991: Terrain elevation and sensor altitude dependence in a semi-analytical atmospheric code. *Canadian Journal of Remote Sensing*, 17, pp. 36–44.

Thomas. R.H., 1993: Ice sheets. In *Atlas of Satellite Observations Related to Global Change*. Edited by R.J. Gurney, J.L. Foster, and C.L. Parkinson. Cambridge University Press, pp. 385-400.

Thomas, D.R. and D.A. Rothrock, 1993: The Arctic Ocean ice balance: A Kalman smoother estimate. *Journal of Geophysical Research*, 98C, pp. 10053-10067.

Thomas, R.H., S.N. Stephenson, R.A. Bindschadler, S. Shabtaie, and C.R. Bentley, 1988: Thinning and ground-line retreat on Ross Ice Shelf, Antarctica. *Annals of Glaciology*, 11, pp. 165–172.

Travis, L.D., 1992: Remote sensing of aerosols with the Earth Observing Scanning Polarimeter. *SPIE Conference Proceedings Polarization and Remote Sensing*, Vol. 1747, pp. 154–164.

Tsu, H. and A.B. Kahle, 1992: Overview of ASTER project in EOS program. *Proceedings of IGARSS '93*, pp. 117–119.

Uchino, O. and I. Tabata, 1991: Mobile lidar for simultaneous measurements of ozone, aerosols, and temperature in the stratosphere. *Applied Optics*, 30, pp. 2005–2012.

Unninayar, S. and K. Bergman, 1993: *Modeling the Earth System in the Mission to Planet Earth Era*, National Aeronautics and Space Administration, Washington, D.C., 133 pp.

Warner, T.T., Y.-H. Kuo, J.D. Doyle, J. Dudhia, D.R. Stauffer, and N.L. Seaman, 1992: Nonhydrostatic, mesobeta-scale, real data simulations with the Penn State University/National Center for Atmospheric Research Mesoscale Model. *Meteorology and Atmospheric Physics*, 49, pp. 209–227.

Waters, J.W., 1993: Microwave limb sounding. In *Atmospheric Remote Sensing by Microwave Radiometry*. Edited by M.A. Janssen. New York, John Wiley & Sons, chapter 8.

Waters, J.W., L. Froidevaux, W.G. Read, G.L. Manney, L.S. Elson, D.A. Flower, R.F. Jarnot, and R.S. Harwood, 1993: Stratospheric ClO and ozone from the Microwave Limb Sounder on the Upper Atmosphere Research Satellite. *Nature*, 362, pp. 597–602.

Wessman, C.A., 1993: Towards a unification of forest carbon budgets. *Bulletin of the Ecological Society of America*, 74, pp. 45–47.

Wielicki, B. A. and B. R. Barkstrom, 1991: Clouds and the Earth's Radiant Energy System (CERES): An Earth Observing System experiment. *AMS Second Symposium on Global Change Studies*, New Orleans, LA, Jan. 14-18, p. 11–16.

Willson, R.C. and H.S. Hudson, 1991: The sun's luminosity over a complete solar cycle. *Nature*, 351, pp. 42–44.

Wilson, S. (ed.), 1990: *EOS: A Mission to Planet Earth*, Internetwork, Del Mar, California, 36 pp.

Xinmei, H., T.J. Lyons, R.C.G. Smith, J.M. Hacker, and P. Schwerdtfeger, 1993: Estimation of surface energy balance from radiant surface temperature and NOAA AVHRR sensor reflectances over agricultural and native vegetation. *Journal of Applied Meteorology*, 32, pp. 1441–1449.

Yamaguchi, Y., H. Tsu, and H. Fujisada, 1993: Scientific basis of ASTER instrument design. *SPIE Proceedings*, 1939, pp. 150–160.

Yarnal, B. and J.D. Draves, 1993: A synoptic climatology of stream flow and acidity. *Climate Research*, 2, pp. 193–202.

3 1836 0015 4825 4